APPLIED CONIC FINANCE

This book introduces the brand-new theory of conic finance, also referred to as the two-price theory, which determines bid and ask prices in a consistent and fundamentally motivated manner. While theories of one price classically eliminate all risk, the concept of acceptable risks is critical to the foundations of the two-price theory, which sees risk elimination as typically unattainable in a modern financial economy.

Practical examples and case studies provide the reader with a comprehensive introduction to the fundamentals of the theory, a variety of advanced quantitative models, and numerous real-world applications including portfolio theory, option positioning, hedging and trading contexts.

This book offers a quantitative and practical approach for readers familiar with the basics of mathematical finance to allow them to boldly go where no quant has gone before.

DILIP MADAN is Professor of Mathematical Finance at the Robert H. Smith School of Business. He currently serves as a consultant to Morgan Stanley, Norges Bank Investment Management and MarketToppers. He has also consulted with Citigroup, Bloomberg, the FDIC, Wachovia Securities, Caspian Capital and Meru Capital. He is a founding member and past President of the Bachelier Finance Society.

WIM SCHOUTENS is a Professor in Financial Engineering at the KU Leuven – University of Leuven in Belgium. He has extensive practical experience of model implementation and is well known for his consulting work to the banking industry and other institutions. Schoutens has authored several books on a variety of financial engineering-related topics such as Lévy processes, credit risk and contingent capital. He is also Managing Editor of the *International Journal of Theoretical and Applied Finance* and *Quantitative Finance*, and Associate Editor of both *Mathematical Finance* and *Review of Derivatives Research*.

APPLIED CONIC FINANCE

DILIP MADAN
University of Maryland, College Park

WIM SCHOUTENS
KU Leuven – University of Leuven

CAMBRIDGE
UNIVERSITY PRESS

CAMBRIDGE
UNIVERSITY PRESS

University Printing House, Cambridge CB2 8BS, United Kingdom

Cambridge University Press is part of the University of Cambridge.

It furthers the University's mission by disseminating knowledge in the pursuit of education, learning, and research at the highest international levels of excellence.

www.cambridge.org
Information on this title: www.cambridge.org/9781107151697

First published 2016

Printed in the United Kingdom by Clays, St Ives plc

A catalogue record for this publication is available from the British Library.

ISBN 978-1-107-15169-7 Hardback

To our perennial supporters:

to Vimla, Maneka and Shivali
—Dilip

to Ethel, Jente and Maitzanne
—Wim

Contents

Preface

All models are wrong, but some are a bit more useful than others.

In traditional financial mathematics the focus of derivative pricing is often solely on the so-called risk-neutral price (cfr. the law of one price), or the (equilibrium) price at which we supposedly can buy and sell. However, in real markets, one observes continuously two prices, namely the price at which the market is willing to buy (bid) and a price at which the market is willing to sell (ask). Hence, if one is going for an immediate transaction, one should take into account the direction of trade (buying or selling). This book presents a theory, referred to as the two-price theory or the conic finance theory (see later for the explanation of the word "conic"), which is about determining such bid and ask prices in a consistent and fundamentally motivated manner.

The law of one price, or (just for derivative pricers) the risk-neutral price, is a useful theoretical abstraction that serves many purposes including the concept of marking to fair market values the asset and liability positions of economic participants. Actual markets, however, recognize that the real value for positions will depend on many factors, including trade directions, the size of the trade, how fast it is conducted and who the actual counterparties turn out to be. Now, between the abstraction proposed by the law of one price and the actual realization of value in transaction lies the abstraction of the two-price theory. It provides some advantages by enabling the marking of positions to be done conservatively to higher levels (ask) for liabilities and lower ones (bid) for assets under unfavourable immediate unwinds.

Critical to the foundations of the two-price theory is an underlying concept of acceptable risks in the economy. Classically, in theories based on the law of one price, one often introduces the concept of complete markets that deliver unique prices while simultaneously eliminating all risk. Acceptable risk is then no longer an issue. The basic foundations of two-price theory is the recognition that risk elimination is typically unattainable and not available. Hence, acceptable risks must be defined.

We therefore start by recognizing that in a modern financial economy all risks cannot be eliminated and risk exposures must be tolerated. Hence, the set of acceptable risks must be defined as a financial primitive of the financial economy.

First, let us consider supposing that the non-negative random variables are acceptable at zero initial cost. Such variables should actually always be acceptable since they are in fact arbitrages: you get a non-negative, potentially positive cash flow for zero cost.

In a second step, let us elaborate on the traditional risk-neutral valuation of risks. In that traditional risk-neutral setting, the price, or value, of a financial derivative equal to the discounted risk-neutral expectation of the payoff under the risk-neutral measure. Assuming that the risk-neutral price is the "correct" price to trade on, the market will "accept", at initial zero cost, risks of non-negative risk-neutral value.

However, in reality one may recognize that the market does not trade at the risk-neutral price and that the price at which one can (immediately) trade depends on the direction of the trade, and there are typically current bid and ask quotes at which one may sell or buy. We are thus led to considering convex proper subsets of the set of risks of non-negative risk-neutral value (which forms a half-space), but containing the non-negative random variables as models for the set of potentially acceptable zero-cost cash flows by markets.

When the set of acceptable loss exposures is modelled by such a convex set of random variables that contains the always acceptable non-negative random variables, these exposures turn out to be just those that have a positive expectation, not just under the risk-neutral measure but under a whole set of different probability measures that may be called scenario probabilities. The set of acceptable loss exposures is then a convex cone of random variables, in the sense that it is closed under scaling.

The best price at which one may then sell a random outcome of cash flows is then the infimum or minimal expectation of the cash flow being priced under all the scenario probabilities. Similarly, the best price at which one may buy a random outcome of cash flows is the supremum or maximal expectation under all the scenario probabilities. These two extremal expectations are generally different, with the lower one becoming the bid price of a two-price economy while the higher one is the ask price. The former applies to assets that may eventually be sold while the latter applies to liabilities that must eventually be bought and delivered. By virtue of the infimum, the bid price is a concave function on the space of random outcomes and is suited to being maximized. The ask price, on the other hand, is a convex function of the random outcomes, suited to being minimized. In this way we obtain new market-based objectives for financial decision making. The book presents applications of these objectives to many financial decision problems, be they the design of portfolios, the valuation of credit exposure, the construction of hedges or the optimal design of derivative portfolios.

The applications require one to be quite specific about the set of scenario probabilities. Here the book makes direct connections to the literature on expectation with respect to non-additive probability that goes back to the work of Choquet. We essentially recognize the flaw embedded in expectations of treating all probabilities with an equal status, knowing full well that our uncertainty about such probabilities for extreme and rare tail events on either side of the gain–loss spectrum can be significant. Non-additive probability allows one to conservatively inflate the probabilities of large loss outcomes while simultaneously deflating those associated with large gains. When the cone of acceptable risks is defined purely by the risk distribution function, and in addition we ask for bid/ask prices of comonotone risks to be additive, it turns out that the bid/ask price is an expectation with respect to non-additive probability exactly as defined by Choquet. The book then brings to bear on financial problems the generalized Choquet expectation as the right objective for such decision making.

The context for the absence of risk elimination is essentially one of exposure to a multitude of surprise movements in markets of various directions and sizes. The complete markets context of continuous price processes eliminates such risks by assuming continuity and no surprise moves. Continuity in price processes is actually impossible, as it would require a continuum of trades at all price points between two price points at which one has a trade. With a finite and possibly large number of trades occurring over an interval of time, one typically registers a number of jumps in prices up or down that are better modelled by pure jump processes, and this requires a working knowledge of Lévy processes, their analytics and simulation. Chapters 2 and 3 provide this background. Chapter 1 introduces the two prices as they are observed in option markets and the no-arbitrage implications for them.

Conic finance is then presented in Chapter 4, with its link to Choquet expectations. Chapter 5 introduces the mechanics of conic pricing in the classical contexts of trees, Brownian motion risks and other processes. The book then takes on, in turn, the conic pricing of credit risk, design of portfolios, hedging of financial and insurance risks, the design of derivative positions and the trading of Markov processes.

Importantly, the conic pricing of credit risk shows how counterparty credit risk, when conservatively valued at the bid price, results in larger markdowns than would occur under risk-neutral pricing. On the other hand, when it comes to the debt valuation adjustment, since it is a liability, it must be priced at the ask. This mitigates and can even eliminate this adjustment.

Conic portfolio theory teaches us that the risk in a demeaned portfolio is the cost of acquiring its negative to get back to zero. Hence, the right measure for risk to offset the reward related to the mean return is the ask price for the return on the negative demeaned portfolio. The risk–reward frontier is then between the mean and this ask price. Both are measured in dollars and trade one to one. One may contrast with mean variance theory that measures reward in dollars and risk in squared dollars and then attempts unrealistically to compensate with a trade-off coefficient.

With respect to hedging liabilities, we gain the ability, in a conic setting, to compare different hedges, with the better ones being those that reduce the ask price for the hedged liability. By contrast, in a complete markets setting, two hedges that both eliminate risk are equivalent. Since, in reality, one is often unlikely to be able to eliminate risk, it is important to know which are the better hedges.

The design of derivative positions from a conic perspective leads to the selection of distribution functions or their inverses. Examples illustrate a variety of ways of organizing such distribution functions in closed form. The solutions involve the interactions between the physical and risk-neutral distribution functions.

In the final chapter, we show how the conic finance machinery can be put to work to construct optimal trading policies. We work in a discrete time Markovian context. It turns out to be critical here that the valuation is conducted by an expectation with respect to non-additive probability, for, with a classical conditional expectation operator, both the value and policy iterations embedded in the associated Markov decision process fail.

Acknowledgements

This book is the work of the authors, but without the support of our colleagues, PhD students, co-authors, people we met on seminars and conferences, and referees of the papers we published, it would not have been possible.

We would like to thank explicitly Peter Carr, José Manuel Corcuera, Alexander Cherny, Samuel Cohen, Jan De Spiegeleer, Ernst Eberlein, Robert Elliott, Monika Forys, Florence Guillaume, Philip Heckman, Ali Hirsa, Pankaj Khandelwal, Ajay Khanna, Peter Leoni, Ine Marquet, Martijn Pistorius, Yazid Sharaiha, Mitja Stadje, Juan Jorge Vicente Alvarez, King Wang, Shaun Wang and Marc Yor.

Our gratitude and respect also go to the team of Cambridge University Press for the enthusiasm and professionalism with which they have embraced this book project.

1

Financial Mathematics Principles

1.1 Financial Derivatives and Derivatives Markets

A financial derivative is a special type of financial contract whose value and payouts depend on the performance of a more fundamental underlying asset. One finds derivatives on the basis of all kinds of underlying entities, such as equity derivatives, whose performance is linked to the behaviour of underlying stock prices or stock indices; fixed income derivatives, whose payout depends on the level of interest rates; currency derivatives, that are connected with one or more FX exchange rates; derivatives on commodities which, for example, can depend on the (joint) evolution of oil, gas, gold, orange juice or any other commodity prices. Actually, they can come in all forms and on the basis of all kinds of underliers; sometimes even a combination of several underliers from different asset classes (hybrids).

Derivatives come into existence in modern economies and encourage price discovery in free markets with consequent price volatility. Often, good business planning requires some limited price stability. This can be (partially) provided by derivatives. One then shifts price risk to professionals better positioned to manage the oscillations. Derivatives can be used for many purposes. They can be used not only to mitigate price risk but also to speculate on it. Many contracts and structured products are implemented using derivatives. A sophisticated risk-management of a portfolio uses derivatives to hedge away as completely as possible all undesired exposure. Some capital instruments even have derivative features and can be seen essentially as cash flows being made contingent on the resolution of future uncertainties.

Derivatives are traded on exchanges (like the CBOE) or over the counter (OTC). Derivatives traded on exchanges are typically standardized; OTC derivatives are tailor made. You can compare it with a clothes collection that comes in one design and is available in standardized sizes (XS, S, M, L, XL, ...) and colours. These you find in retail shops (compare with exchanges). If, however, you desire a different design, which fits your frame and specific body form and is printed in your preferred (non-standard) colour, you have to go to a tailor and have it custom made (compare with OTC).

Both forms have their advantages and disadvantages. You buy the standardized contract via the exchange and you neither know nor care about the actual counterparty who is selling the derivative to you (this is like knowing the shop but not the person/tailor who actually made your clothes). It is the exchange that takes care that the obligations in the contract are met and the exchange is taking the risk of counterparty default while simultaneously putting in place measures against such default. As long as the exchange itself doesn't default, the contract is honoured, as far as you are concerned. This is different with OTC derivatives; here there is no intermediator and you are dealing directly with your counterparty (this is like interacting with the tailor who is actually making your clothes). A default of this counterparty can lead to huge losses since, no matter what was agreed, a default of your counterpart can mean non-delivery of what has been promised. Of course, OTC derivatives can be designed to your very specific needs and may better suit you than a standardized exchange-traded product. On the other side, exchange-traded products are typically more liquid and can be bought and sold back easily; there are continuously bid and ask prices quoted at which you can immediately transact. The bid price is the price at which you can sell: the price that somebody is bidding for the asset. The ask price is the price at which you can buy: the price that somebody is asking for it. The more common the product is, the lower the spread, i.e. the difference between the ask and the bid price. It is much harder to unwind or sell back an OTC derivative. It can actually involve a new negotiation round with unclear terms of settlement. Maybe your counterparty is not even willing to take it back, and even in case that it is, the spread is usually much higher than the spread of exchange-traded products because the product is very specific and exotic. It involves many more uncertainties (model risk, calibration risk, ...), its hedging is more complex (and hence it is also more involved for your counterparty to unwind it) and more (safety) margin is charged. It is just a deal between a very limited number of parties, and therefore the ask and demand forces are not in place as they would be on exchanges open to everybody on the globe.

Derivatives are omnipresent in today's financial markets. There are various types. One has futures and forwards, which are basically contracts to buy or sell the underlying at a predetermined price in the future on a predetermined date. Hence the limited price stability offered. One also has swaps, that agree on exchanging certain uncertain cash flows over a predetermined period, and finally one has options, that are agreements in which the holder has a right to buy or sell the underlying at specific conditions. To buy an option, one has to pay the price or the premium for the option; swaps are often initiated at zero cost and have therefore an initial market price equal to zero. However, they can be written on a huge underlying notional and, although of zero value when the deal is struck, can bear significant risks. The actual size of derivatives' markets is not easy to estimate, but at the end of 2013, the Bank for International Settlements (BIS) estimated the total notional outstanding for OTC derivatives at USD 710 trillion and at a USD 18.6 trillion gross market value.

The theory we develop in this book, on how to determine the bid and ask prices and how to deal with the risks of such derivatives, is applicable to all derivative types over all asset classes mentioned. However, we mainly focus on equity derivative options. The basic examples are European Call and Put options, often referred to as vanilla options because they can be regarded as the most simple type of options. Before we define them and illustrate their use in Section 1.3, we first recall the concept of the risk-free bank account.

1.2 The Risk-Free Bank Account and Discount Factors

A risk-free interest rate can be viewed as the interest rate that rewards the depositor of some amount of money on lending it to a counterparty that cannot default. Traditionally, "risk-free" interest rates were derived from the rates associated with Treasury bonds issued by governments. However, nothing is actually without any risk of default, and history has shown that some governments can and have defaulted on their obligations (Argentina, Greece, Cyprus, ...).

If you deposit USD 1000 today in a risk-free account earning an interest rate of $r = 2\%$ (per annum), then the value in one year's time would be USD 1020 $(= 1000(1 + 0.02))$. In this calculation we have assumed that the interest is compounded annually.

Now assume that you want to have exactly USD 2000 (N) in the account in three years' time (T) from now, and assume we have a (flat) interest rate of $r = 2\%$ (per annum). How much do you have to put into the risk-free account now? The answer is given by the formula:

$$\frac{N}{(1+r)^T} = \frac{2000}{(1+0.02)^3} = 1884.64.$$

Indeed, after putting USD 1884.64 into the account, it grows under a 2% interest rate after one year to USD 1922.33 $(= 1884.64(1+0.02))$ and after another year to USD 1960.78 $(= 1922.33(1 + 0.02))$ and finally, after the last year, to USD 2000 $(= 1960.78(1 + 0.02))$. The interest is compounded again annually, and we call USD 1884.64 the present value of USD 2000 received in three years. The ratio of both is called the discount factor or the price of future money and is equal to

$$\frac{1}{(1+r)^T}.$$

In our example it equals 0.9423. It is basically today's value of receiving USD 1 in T years from now.

In the above examples, we had annual compounding; i.e. interest is paid after each year. One could also have other schemes of interest payment, such as semi-annually, quarterly, monthly, weekly and daily. In general, then, N units of currency will grow to

$$N\left(1+\frac{r}{m}\right)^{mT}$$

if we compound m times per year and keep the money in our account until time T. The discount factor then equals

$$\left(1+\frac{r}{m}\right)^{-mT}.$$

For example, if we have quarterly compounding, $m = 4$ and $r = 2\%$, and we again put 1000 on the account for one year, the account grows after a first interest payment to 1005 ($= 1000(1 + 0.005)$) after three months. After three other interest payments, one at six months, one at nine months and one after a year, it would accumulate at the end of year one to 1020.15 ($= 1000(1 + 0.02/4)^4$), which is just slightly more than with annual compounding. The obvious reason is that, after an interest payment is made, the investor starts to earn interest on this payment. Note that different compounding conventions lead to different rate quotes consistent with the same price for future money at a fixed future date.

When pricing derivatives, we usually assume continuous compounding of interest rates, meaning that compounding (i.e. receiving interest) occurs continuously, i.e. over an infinitesimally small period of time, or in other words if $m \to \infty$. Our discount factor then becomes

$$\lim_{m\to\infty}\left(1+\frac{r}{m}\right)^{-mT} = \exp(-rT).$$

In Figure 1.1, one clearly sees the convergence if $m \to \infty$; one can also see the differences between yearly ($m = 1$), semi-annually ($m = 2$), quarterly ($m = 4$), monthly ($m = 12$) and weekly ($m = 52$) compounding on an investment of USD 1000 at $r = 2\%$ during a period of exactly one year ($T = 1$).

In reality, each maturity T and quoting convention has its own interest rate, $r(T)$ say, reflecting the market expectation of changing interest rates over the given period. We then speak about a yield curve and the interest rate term structure.

The risk-free yield curve is a curve showing several yields or risk-free interest rates across different contract lengths (maturities), known as the "term", for a risk-free debt contract. One has different curves for different currencies. For example,

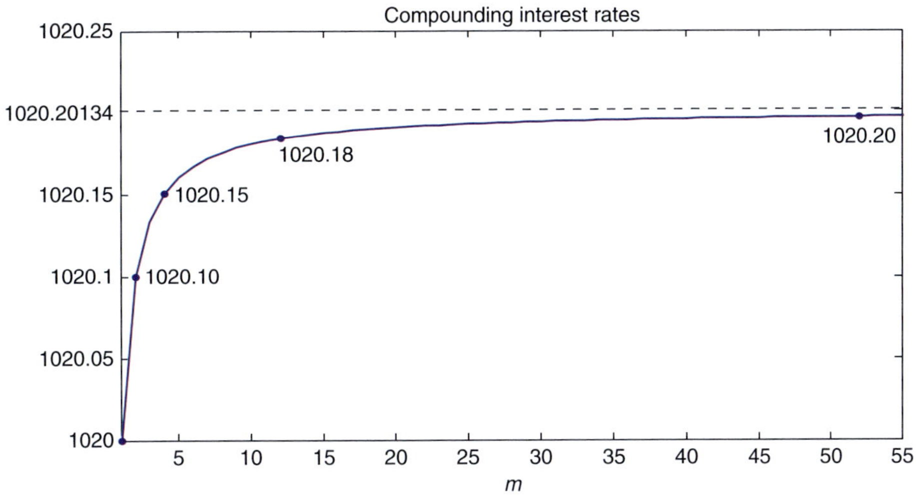

Figure 1.1 Compounding interest rates ($T = 1, N = 1000, r = 2\%, m = 1, 2, 4, 12, 52$)

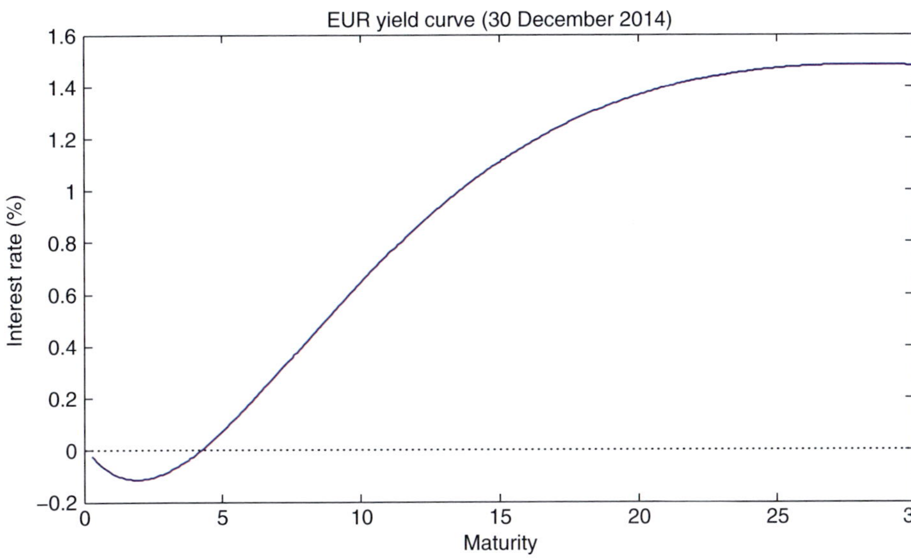

Figure 1.2 EUR yield curve (30 December 2014)

one has the US dollar yield curve based on interest rates paid on US Treasury securities for various maturities, which are assumed to be (almost or as close as possible to) risk-free. In Figure 1.2 the EUR yield curve is shown as of 30 December 2014; this curve is calculated by the ECB and based on "AAA-rated" Euro-area central government bonds.

Interest rate curves are typically upward-sloping, with shorter-term interest rates lower than longer-term interest rates, but can also be downward-sloping (inverted) or humped. An inverted curve, for example, would indicate that the market expects lower interest rates in the future. As seen in the EUR-curve of Figure 1.2, yields can be negative, meaning investors are basically paying money to park their investments in the underlying securities.

Related to a yield curve is a discount curve. The discount curve basically represents the discount factors for the different terms. For a given term, the discount factor is the present value of a currency unit promised at the given term. If $r(t)$ is the yield (continuously compound) associated with the maturity (or term) t, then the discount factor for that term equals $D(t) = \exp(-r(t)t)$. Figure 1.3 shows the discount factors based on the yields of Figure 1.2. From this one can see that, on 30 December 2014, receiving EUR 100 in 30 years' time from that point would be equivalent to receiving EUR 64.16 on that day, since the 30-year interest rate on 30 December 2014 was $r(30) = 0.0147947$. Similarly, since the one-year interest rate was then $r(1) = -0.0008869$, investors need to pay about EUR 100.09 on 30 December 2014 to receive EUR 100 back on 30 December 2015.

Discount factors are used to discount cash flows (at the risk-free rate). Assume, for example, a cash flow consisting of N payments of EUR C_i paid out at times t_i, $i = 1, \ldots, N$. Then the present value (PV) of this cash flow equals:

$$PV = \sum_{i=1}^{N} C_i \exp(-r(t_i)t_i) = \sum_{i=1}^{N} C_i D(t_i),$$

with $D(t)$ denoting here the EUR-related discount factor with term t.

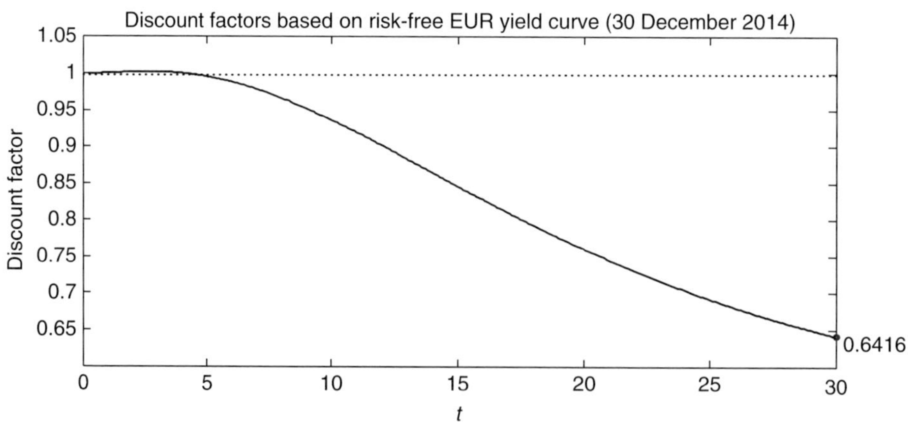

Figure 1.3 Discount factors (EUR, 30 December 2014)

1.3 Vanilla Options

A *European Call option* gives the right to the holder (the buyer) of the option to buy from the writer of the option (the seller) at a predetermined future time point $t = T$, called the maturity, the underlying, a stock, S say, for a predetermined price K, called the strike. A *European Put option* is similar, but it gives the right to the holder of the option to sell the underlying to the writer of the option at a predetermined future time point $t = T$ (maturity) for a predetermined price K (strike). For such a right the buyer pays a premium to the option seller, when the deal is initiated, say at $t = 0$. We refer to the instrument as well as its premium by the notation $EC(K, T)$ and $EP(K, T)$ respectively. It should be clear from the context whether we are referring to the instrument or the premium.

Besides these so-called *European* options, *American* Call and Put options also exist. These are different from their European counterparts in the sense that the holder can exercise his rights not only at the maturity of the option but during the entire lifetime of the option.

If an option is traded on an exchange, typically bid and ask prices are given continuously. We use the notation $bidEC(K, T)$ and $askEC(K, T)$ for the European Call's bid and ask prices and correspondingly $bidEP(K, T)$ and $askEP(K, T)$ for the European Put. It should be clear from the context whether we are referring to the initial time $t = 0$ price or to a general time t price, with $0 \leq t \leq T$.

The holder of a call option will exercise his right to buy the underlying via the option contract only if this is beneficial to him. This happens when the underlying at maturity ($t = T$) has a market price $S(T)$ that is greater than the strike price: $S(T) > K$. The payoff of the option is then strictly positive and equals the difference $S(T) - K$. In the other situation, it would not be rational to pay via the option the strike price K, which is more than the price $S(T)$ one pays in the market, and the option contract expires, worthless. The payoff is then zero. Summarizing, the payoff of the call option can in general be given by

$$\text{payoff of } EC(K, T) : \max(S(T) - K, 0) = (S(T) - K)^+.$$

Similar reasoning can be given for the put option, where the holder will actually only sell the underlying via the option contract when at maturity ($t = T$) its actual market price $S(T)$ is lower than the strike price K: $S(T) < K$. The payoff of the put equals in general

$$\text{payoff of } EP(K, T) : \max(K - S(T), 0) = (K - S(T))^+.$$

In Figures 1.4 and 1.5 the payoff functions of a European Call and a European Put, respectively, are visualized.

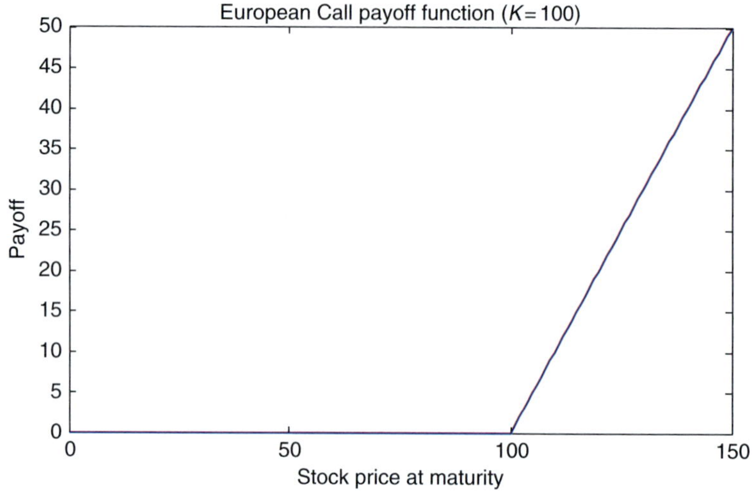

Figure 1.4 European Call (EC) payoff function ($K = 100$)

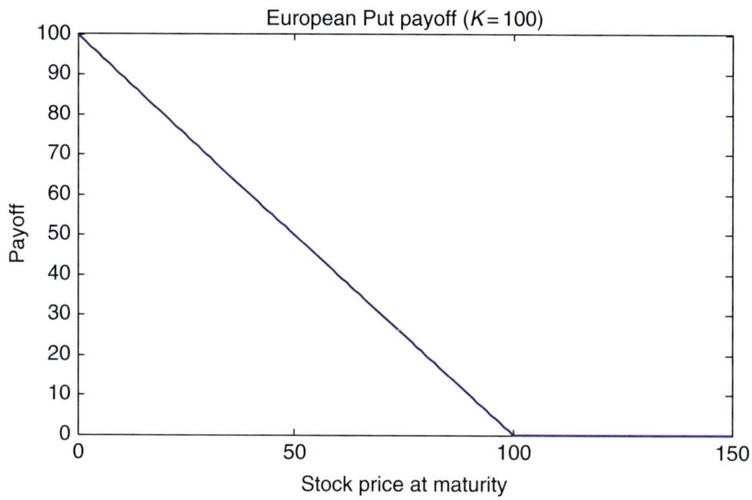

Figure 1.5 European Put (EP) payoff function ($K = 100$)

We say that a European Call option is *out of the money (OTM)* if the current stock price is below the strike. If it is above the strike we say it is *in the money (ITM)*. When the underlying stock prices equals (or is very close to) the strike we say it is *at the money (ATM)*. For a European Put, *out of the money (OTM)* corresponds to the situation where the current stock price is above the strike. If it is below the strike we say it is *in the money (ITM)*. When its stock price equals (or is very close to) the strike we again say it is *at the money (ATM)*. OTM basically means that if one were now to exercise the option (which is actually not allowed in reality for European options, but one assumes one does), the payoff would be zero. ITM means we would then have a non-zero payoff. The amount that one would get if the option were now to be exercised is called the *intrinsic value* of the option. OTM options have a zero intrinsic value. The difference between the current price of the option and its intrinsic value is often referred to as the *time value*: the extra value the option carries because there is still time left until expiry/maturity and the underlying asset can still move in a beneficial direction for the option holder. OTM options have only time value.

Put and call options can be used for different purposes. We illustrate this in the next examples. In the first example, we show how a call option can be used for speculation. The next example shows how a put can be used as a kind of insurance against downward market movements. The third example shows how a call can be used as building block in a very common structured product, a Principal Protected Note (PPN).

Example 1.1 *An investor has USD 10,000 at his disposal and is clearly convinced the stock S will rally in the next year. The stock trades now at USD 50. Vanilla options on the stock are trading as well. The one-year at-the-money (ATM), i.e. with strike equal to the current stock price, European Call option has a bid price of USD 3.85 and an ask price of USD 4.00. With his USD 10,000, the investor can hence buy 200 stocks, or he can buy 2500 European ATM call options (or he could buy some of both). He decides to buy 2500 ATM calls. Note that he is buying and hence has to pay the ask price of USD 4.00. After six months the stock has indeed rallied to USD 65. The call options he bought are now deep in the money, i.e. the current stock price is higher than the strike. They are also closer to maturity (six months) and one call trades at a bid equal to USD 18.00 and has an ask price of USD 18.25. He decides to close his position and sells his 2500 calls at the bid price of USD 18.00. He cashes USD 45,000 and actually makes a 350% profit. Compared with a direct investment in the stock this is much better, since that would have given him only a 30% return.*

Derivatives can be used for speculation and can lead to hugely leveraged positions and hence huge gains. Of course, there is another side to the story. If the stock had not moved higher and, for example, closed after one year at the same level of USD 50, the investment in call options would have led to a 100% loss, since all the call options would have been worthless at maturity. The direct investment in the stock would, in that case, have ended flat and would have shown no loss.

Example 1.2 *At the beginning of a new year an investor steps into the equity market and buys 1000 stocks at a price of USD 65 each. He furthermore has a cash account of USD 2000. His initial total wealth is USD 67,000. After nine months,* *the stock trades at USD 80. He doesn't want to exit his position because he still believes there is upwards potential in the stock. However, he is also worried about the downside. During the first nine months of the year the stock has been rallying nicely (at the beginning of the year he entered at USD 65!) and it is now vulnerable to potential negative market sentiment in the next months until year-end. He wants to close his year positively and seeks protection against adverse downside market movement without exiting his position and potentially missing a continuation of a rally in the stock. Derivatives are trading on the underlying stock. An out-of-the-money (OTM), i.e. with strike below the current spot, European Put with maturity of three months with strike USD 70 is trading with a bid of USD 1.80 and an ask of USD 2.00. He decides to spend his USD 2000 cash to buy protection and buys (at the ask price) 1000 three-month European Puts with strike USD 70. By year end the market indeed went down and the stock is now trading at USD 60. Without any protection (and assuming no interest payments on his cash account), he would have been down for the year by 7.46% (USD 5000/USD 67,000). However, due to his put options, he receives an additional payoff. At expiration the put options gave him a payoff of USD 10 each. He thus received USD 10,000. Compared with the value of his position in the beginning of the year (USD 65,000+USD 2000), he now has stock worth USD 60,000 and USD 10,000 in cash, or USD 70,000 in total. He is therefore up 4.48% (USD 3000/USD 67,000) for the year.*

Example 1.3 *Salespeople have learnt that their retail customers are very interested in investing their money in the medium term in the stock market, especially in some new social media companies. However they are also worried about losing their investment. The structuring team therefore designs the following structured product: a Principal Protected Note (PPN). For each USD 1000 you invest, you receive after four years your initial investment (USD 1000) plus 60% of the positive performance of a social media stock S. If the stock S were to end after four years below its value at initiation, the investor would still get his initial investment back. Denoting with $S(0)$ the initial stock price of S and with $S(4)$ the stock price after four years, the investor hence receives:*

$$PPN = 1000 + 1000 \times 60\% \times \max\left(\frac{S(4) - S(0)}{S(0)}, 0\right). \tag{1.1}$$

The final wealth with an initial investment of USD 1000 in either the PPN or in the stock is compared in Figure 1.6.

Assume that investing now USD 0.82 into a risk-free account would give in four years' time USD 1; this corresponds approximately with an interest rate of 5%. Assume $S(0) = 20$. Note that in this setting Equation (1.1) becomes

$$PPN = 1000 + 30 \times \max\left(S(4) - 20, 0\right) = 1000 + 30 \times \left(S(4) - 20\right)^+ .$$

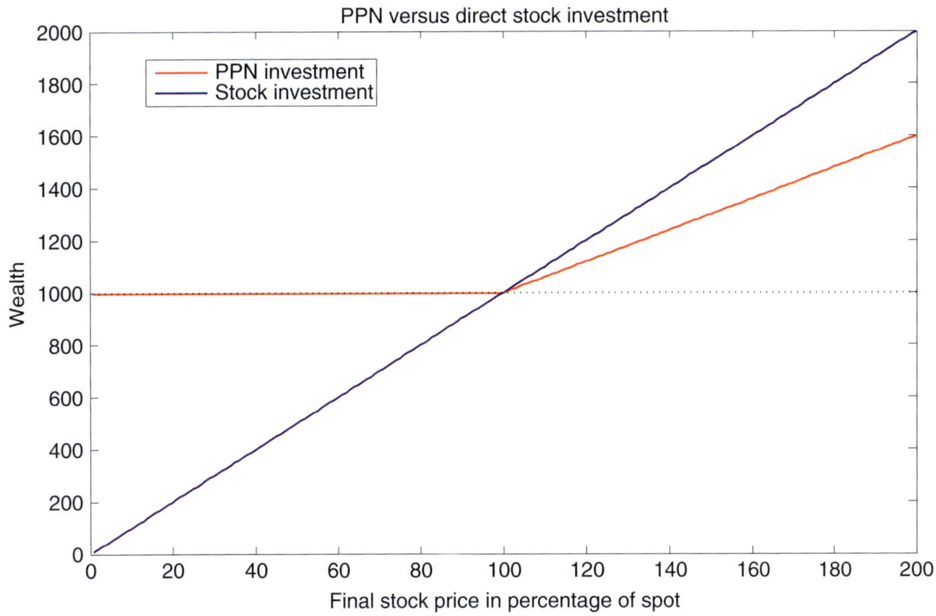

Figure 1.6 Comparing the Principal Protected Note (PPN) with a direct stock investment. Initial investment is USD 1000; participation rate equals 60%.

Further assume that an at-the-money (ATM), i.e. with strike $K = 20$*, four-year European Call option is available and trades at bid USD 5.25 and ask USD 5.50. The bank implements this product therefore as follows. For each USD 1000 it receives, it puts USD 820 on the risk-free bank account. It also buys 30 ATM call options at the ask price of USD 5.50 at a total cost of USD 165. From the original USD 1000, the bank has hence put USD 820 in a risk-free account and has bought for USD 165 call options; it still has USD 15 (or 1.5% of the investment amount) left which it considers as its margin. After four years the amount in the bank account has grown to USD 1000; furthermore, the 30 call options expire and have each as payoff* $(S(4)-20)^+$*. The total hence equals* $1000+30\times\max(S(4)-20, 0)$*, exactly the amount it owes the structured product investor.*

1.4 The No-Arbitrage Assumption

The main assumption in the financial engineering theory for pricing financial derivatives is the absence of arbitrage. Arbitrage is the possibility of making money without any risk and without any starting capital. We assume that arbitrage is not possible. This means that if you can enter into a position which pays you in the future, say at time T, a payout $X \geq 0$ and in some situations (with non-zero probability) a strictly positive payout $X > 0$, then the price to enter (at any time before) for it needs to be strictly positive as well. Indeed, imagine you can enter at a zero price or at a strictly negative price, meaning you are paid to enter into the position, then you run no risk, since you can't lose ($X \geq 0$) and in some cases you can receive a positive payoff ($X > 0$). In addition, you even received in the case that the price was strictly negative some premium at initiation. This is an arbitrage and is ruled out. Hence the price to enter the position needs to be strictly positive.

If we work in such a framework with bid and ask prices, no arbitrage means that the price for **buying** X needs to be strictly positive. This implies that the ask price for it needs to be strictly greater than zero.

If for X itself no bid and ask prices are available, but X is a combination of positions for which bid and ask prices are quoted, then the total price equals the sum of the ask prices of the position one needs to buy minus the sum of the bid prices of the component one needs to sell in order to achieve the total position X. Examples of this will be given in the next sections.

1.5 The Put–Call Parity

In this section, we elaborate on a special relation, called the put–call parity, between the price of a European Call and the price of a European Put option on the same underlying (non-dividend-paying) stock, with current price $S(0)$, say, with the same time to maturity T and the same strike price K. We first start by proving the put–call parity in the absences of a bid and ask spread, i.e. the prices at which one can buy the options equal the prices at which one can sell. In this one-price world, and in absence of arbitrage, the following relation holds:

$$S(0) - \exp(-rT)K + EP(K,T) - EC(K,T) = 0. \qquad (1.2)$$

This relation can be proven as follows. Assume that $S(0) - \exp(-rT)K + EP(K,T) - EC(K,T) > 0$; then you could do the following: (short) sell the stock, put $\exp(-rT)K$ on the risk-free account, write (i.e. sell) the put option and buy the call option. You receive $S(0)$ for the stock, have to put $\exp(-rT)K$ in the bank account, you receive the put premium $EP(K,T)$ and you have to pay the call premium $EC(K,T)$. You still then have in your hands: $S(0) - \exp(-rT)K + EP(K,T) - EC(K,T)$, which we assume to be strictly positive.

The total payoff at maturity of this position equals $-S(T) + (S(T)-K)^+ - (K - S(T))^+ + K = 0$. Indeed, you have to close your short position in the stock and buy it back ($-S(T)$); you receive the call payoff ($(S(T) - K)^+$) but you have to pay the put payoff ($-(K - S(T))^+$) and your initial position on the bank account has now grown over the period of length T to the exact amount K. Hence, you started with nothing and at the end of the trade there is no risk any more, but after initiating the trade you had some (free) money left. This is an arbitrage opportunity and hence not possible. Our initial assumption $S(0) - \exp(-rT)K + EP(K,T) - EC(K,T) > 0$ hence cannot be true.

Now assume $S(0) - \exp(-rT)K + EP(K,T) - EC(K,T) < 0$; then you could do the opposite: buy the stock, borrow $\exp(-rT)K$ from the bank, write the call option and buy the put option. Then, at time zero, you have to pay $S(0)$ for the stock, cash $\exp(-rT)K$ from the bank account, pay the put premium $EP(K,T)$ and receive the call premium $EC(K,T)$. You still have in your hands: $-S(0) + \exp(-rT)K - EP(K,T) + EC(K,T)$, which we assume to be strictly positive.

The total payoff at maturity of this position equals $S(T) - (S(T) - K)^+ + (K - S(T))^+ - K = 0$. Indeed, at maturity your stock is worth $S(T)$, you receive the put payoff but you have to pay the call payoff. Finally, you have to pay back the borrowed amount plus interest rates, namely the exact amount K. Hence you started with no money and at the end of this trade there is again no risk any more, but after initiating the trade you had some (free) money to spend. This is again an arbitrage opportunity and hence not possible. Our initial assumption $S(0) - \exp(-rT)K + EP(K,T) - EC(K,T) < 0$ hence cannot be true. In conclusion, we must have $S(0) - \exp(-rT)K + EP(K,T) - EC(K,T) = 0$.

The situation is further complicated in the two-price world where the price of derivatives depends on whether you are buying or selling. The first trade above, namely, short selling the stock, putting $\exp(-rT)K$ on the risk-free account, writing the put option and buying the call option, gives you $S(0)$ for the stock; you have, since you are buying, to pay the ask price for the call, $askEC(K,T)$, you sell the put and hence receive the bid price of it $bidEP(K,T)$ and you have to put $\exp(-rT)K$ on the bank account. Net, this leads to: $S(0) - askEC(K,T) + bidEP(K,T) - \exp(-rT)K$, which if strictly positive would lead to an arbitrage. Hence

$$S(0) - askEC(K,T) + bidEP(K,T) - \exp(-rT)K \leq 0$$

and so

$$S(0) - \exp(-rT)K \leq askEC(K,T) - bidEP(K,T).$$

Similarly, for the second trade, namely buying the stock, borrowing $\exp(-rT)K$ from the bank, buying the put option and selling the call option, you have to pay $S(0)$ for the stock, receive the call's bid price $bidEC(K,T)$, pay the put's ask price $askEP(K,T)$ and cash $\exp(-rT)K$ from the bank account. The total amount you have left in your hands after the trade equals: $-S(0) + bidEC(K,T) - askEP(K,T) + \exp(-rT)K$. This amount cannot be strictly positive because otherwise you could have a free lunch! Therefore:

$$S(0) - bidEC(K,T) + askEP(K,T) - \exp(-rT)K \geq 0$$

or

$$S(0) - \exp(-rT)K \geq bidEC(K,T) - askEP(K,T).$$

In conclusion, we have:

$$bidEC(K,T) - askEP(K,T) \leq S(0) - \exp(-rT)K \leq askEC(K,T) - bidEP(K,T).$$

1.6 The Forward Contract

A forward contract is an agreement between two parties: the forward buyer, who is said to be "long the forward", and the forward seller, who is so-called "short the forward". The forward buyer agrees to buy from the forward seller the underlying asset, a non-dividend paying stock, say, at a certain future time point T, the maturity, for a given price K (the strike). The difference with an option is that the forward buyer *must* buy the underlying, rather than just having the right to buy it. Depending on the agreed strike, an up-front premium is paid at initiation from the forward buyer to the seller or the other way around. The payoff of such a forward contract for the buyer equals $S(T) - K$. If no up-front premium is paid and the deal is considered fair as it is, the special strike K^*, for which this is the case, is called *the stock's forward price*.

A forward can be synthetically created out of two vanilla options on the same underlier. Indeed, being long a forward contract is equivalent to buying a European Call option and selling a European Put option on the same underlier, both maturing at time T and both with K as strike, because the payoff equals: $(S(T) - K)^+ - (K - S(T))^+ = S(T) - K$.

In the one-price world, the up-front premium is therefore equal to $EC(K,T) - EP(K,T)$, which, by the put–call parity, equals

$$Forward(K,T) = EC(K,T) - EP(K,T) = S(0) - \exp(-rT)K.$$

The forward price is the strike K^* for which $EC(K^*,T) = EP(K^*,T)$. One can easily see from the put–call parity that this is the case when $K^* = \exp(rT)S(0)$.

Again, the situation in a two-price framework is a bit more involved. If you buy the call at $askEC(K,T)$ and write a put for the price $bidEP(K,T)$, and in addition go short the forward and would receive for it $bidForward(K,T)$, the total money one has left, namely $bidForward(K,T) - askEC(K,T) + bidEP(K,T)$, cannot, due to the no-arbitrage assumption, be strictly positive, since the risks are completely covered. Hence

$$bidForward(K,T) \leq askEC(K,T) - bidEP(K,T).$$

Similarly, if you sell the call and buy the put option – the former is done at bid price $bidEC(K,T)$ and the latter at the ask price $askEP(K,T)$ – and in addition go long, i.e. buy the forward at $askForward(K, T)$, the total cash received, i.e. $bidEC(K, T) - askEP(K, T) - askForward(K, T)$, can for the same no-arbitrage reason not be strictly positive. Therefore

$$askForward(K, T) \geq bidEC(K, T) - askEP(K, T).$$

Later on in this book, we will also show that the ask price of a combination of products has a lower or equal price than the sum of the ask prices of its components and that the bid price is greater than or equal to the sum of the components' bid prices. This is the so-called sub-additivity property; see Section 4.1. This, in combination with a further property of the theory saying that the bid prices of a risk X equals minus the ask price of the risk $-X$ (see Equation (4.7)), leads to:

$$askForward(K, T) \leq askEC(K, T) - bidEP(K, T)$$

and

$$bidForward(K, T) \geq bidEC(K, T) - askEP(K, T).$$

Now we can – in a similar fashion as in the one-price world – determine the special strike for which either the bid or the ask premium is zero. We define *the stock's forward bid price* as the strike, K^*_{bid}, where the bid of the call option matches the ask of the put option: $bidEC(K^*_{bid}, T) = askEP(K^*_{bid}, T)$. Similarly, we have *the stock's forward ask price*, which is the strike, K^*_{ask}, where the ask of the call option matches the bid of the put option; $askEC(K^*_{ask}, T) = bidEP(K^*_{ask}, T)$. Naturally, $K^*_{bid} < K^*_{ask}$, as is shown as well in Figure 1.7.

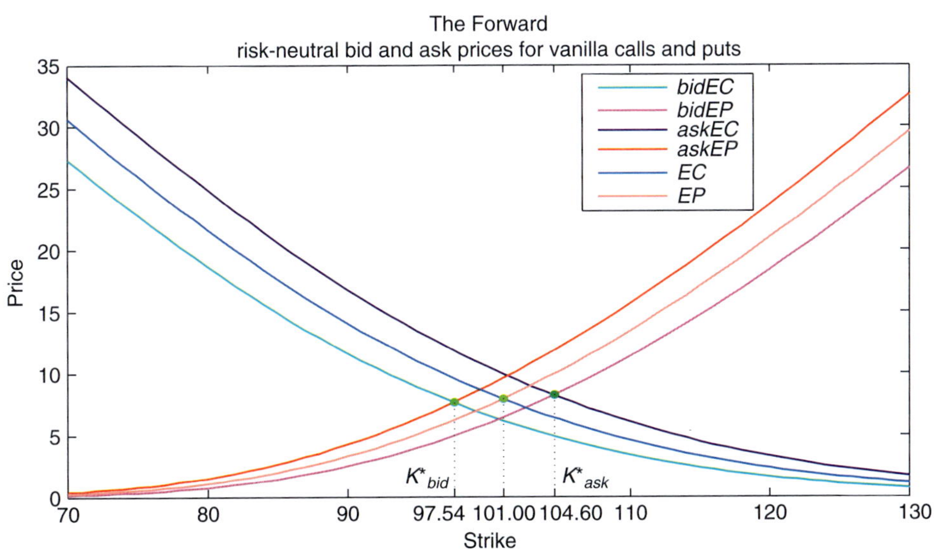

Figure 1.7 The forward contract

1.7 Spread Inequalities

In the world of one price, three inequalities concerning vanilla options play a special role. These are the call and put spread, butterfly spread and calendar spread inequalities. Actually, if for a whole surface of European Calls and Puts on a certain underlier these inequalities are satisfied, they imply that there exists a one-dimensional Markov process consistent with the quoted prices. Or, in other words, there exists an arbitrage-free model driven by a Markov process that produces all these prices (see Carr and Madan, 2005).

In this section we discuss these equalities in the one-price framework and extend these into the two-price framework.

1.7.1 The Call Spread and Put Spread Inequalities

Consider first the call spread inequality. This inequality deals with the prices of two European Call options on the same underlier for two strikes $K_1 < K_2$ and a common maturity T. A call spread is a position long the call with the lower strike and short the call with the higher strike. It hence has as payoff:

$$(S(T) - K_1)^+ - (S(T) - K_2)^+.$$

As can be seen from Figure 1.8, since $K_1 < K_2$, this payoff is always non-negative and has a region where it pays out some strictly positive amount.

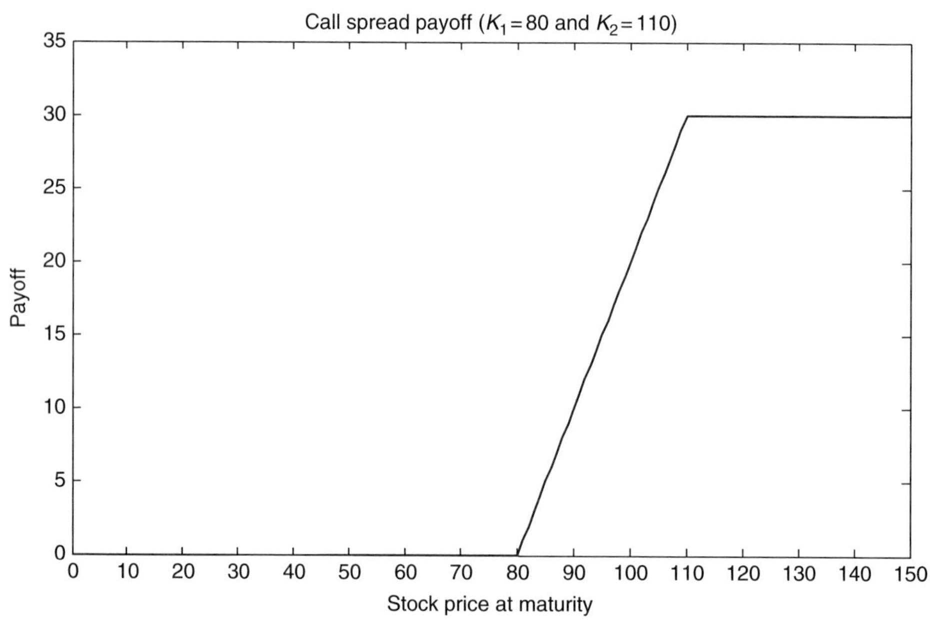

Figure 1.8 Call spread payoff function ($K_1 = 80$ and $K_2 = 110$)

Hence the price for this position at time zero needs to be strictly positive. In other words:

$$EC(K_1, T) > EC(K_2, T), \text{ if } K_1 < K_2.$$

Furthermore, we also have always that $(S(T) - K_1)^+ - (S(T) - K_2)^+ \leq K_2 - K_1$. Therefore,

$$EC(K_1, T) - EC(K_2, T) \leq \exp(-rT)(K_2 - K_1),$$

or

$$0 \leq \exp(-rT)(K_2 - K_1) + EC(K_2, T) - EC(K_1, T).$$

In a two-price framework, entering a call spread can be done by buying the call with strike K_1 at its ask price and selling the call with strike K_2 at its bid price. Hence, all this can be achieved for the total price $askEC(K_1, T) - bidEC(K_2, T)$. Since the position always gives a non-negative payoff and has the possibility to pay out some strictly positive amount, this total amount needs to be strictly positive. Hence

$$askEC(K_1, T) > bidEC(K_2, T), \text{ if } K_1 < K_2.$$

Putting $\exp(-rT)(K_2 - K_1)$ in the risk-free bank account and selling the call with strike K_1 and buying the call with strike K_2 also always gives a non-negative payoff $K_2 - K_1 + (S(T) - K_2)^+ - (S(T) - K_1)^+ \geq 0$. Therefore the price to enter also needs to be non-negative. The cash needed to do this equals $\exp(-rT)(K_2 - K_1)$ (to be put in the bank account) plus $askEC(K_2, T)$ (to buy the call with strike K_2) minus $bidEC(K_1, T)$ (the premium one receives from writing the call with strike K_1). In conclusion,

$$0 \leq \exp(-rT)(K_2 - K_1) + askEC(K_2, T) - bidEC(K_1, T).$$

Similar arguments establish the following inequalities for bid and ask prices for put options, or the put spread:

$$askEP(K_2, T) > bidEP(K_1, T), \text{ if } K_1 < K_2$$

and

$$0 \leq \exp(-rT)(K_2 - K_1) + askEP(K_1, T) - bidEP(K_2, T).$$

1.7.2 The Butterfly Spread Inequality

Consider three strikes $K_1 < K_2 < K_3$ and the purchase of a call struck at K_1 coupled with the sale of $(K_3 - K_1)/(K_3 - K_2)$ calls with strike at K_2 and the purchase of $(K_2 - K_1)/(K_3 - K_2)$ calls with strike K_3. All calls are on the same underlier

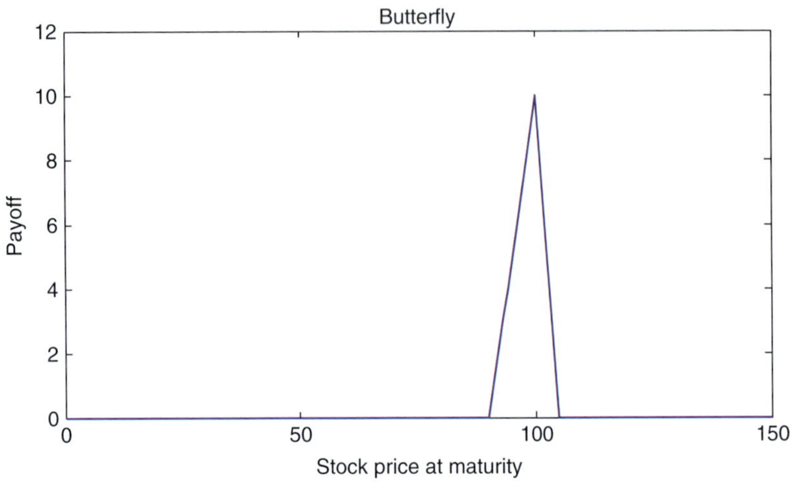

Figure 1.9 Butterfly payoff function ($K_1 = 90$, $K_2 = 100$ and $K_3 = 105$)

and have the same time to maturity T.This is the so-called butterfly spread. The payoff

$$(S(T) - K_1)^+ - \frac{K_3 - K_1}{K_3 - K_2}(S(T) - K_2)^+ + \frac{K_2 - K_1}{K_3 - K_2}(S(T) - K_3)^+ \geq 0$$

is non-negative, as can be seen from Figure 1.9.

Hence in a one-price world we have:

$$EC(K_1, T) - \frac{K_3 - K_1}{K_3 - K_2}EC(K_2, T) + \frac{K_2 - K_1}{K_3 - K_2}EC(K_3, T) \geq 0$$

which one may rewrite as:

$$\frac{EC(K_1, T) - EC(K_2, T)}{K_2 - K_1} \geq \frac{EC(K_2, T) - EC(K_3, T)}{K_3 - K_2}.$$

In a two-price world, we must take into account that we are buying the lowest and highest strike calls and selling the call with strike K_2. Hence, we have that:

$$askEC(K_1, T) - \frac{K_3 - K_1}{K_3 - K_2}bidEC(K_2, T) + \frac{K_2 - K_1}{K_3 - K_2}askEC(K_3, T) \geq 0.$$

Equivalently,

$$\frac{askEC(K_1, T) - bidEC(K_2, T)}{K_2 - K_1} \geq \frac{bidEC(K_2, T) - askEC(K_3, T)}{K_3 - K_2}.$$

For puts, we can obtain similar relations:

$$\frac{EP(K_3, T) - EP(K_2, T)}{K_3 - K_2} \geq \frac{EP(K_2, T) - EP(K_1, T)}{K_2 - K_1}$$

in the one-price world and

$$\frac{askEP(K_3,T) - bidEP(K_2,T)}{K_3 - K_2} \geq \frac{bidEP(K_2,T) - askEP(K_1,T)}{K_2 - K_1}$$

in the two-price world.

1.7.3 The Calendar Spread Inequality

The calendar spread inequality in the one-price world compares the prices of two European Call options on the same underlier but with different maturities $0 < T_1 < T_2$:

$$EC(K\exp(-r(T_2 - T_1)), T_1) \leq EC(K, T_2),$$

where we assume no dividends. To prove this inequality, suppose that this inequality would not hold and

$$EC(K\exp(-r(T_2 - T_1)), T_1) - EC(K, T_2) > 0;$$

then one could buy the long dated call and sell the short dated call and still have, due to the assumed inequality, some positive money left at time zero. One then waits until T_1 and if the shorted dated call would end out of the money, the short dated call is expiring worthless and we still hold the long dated call which can only give us a non-negative payoff. If the short dated call ends in the money, we have a negative payoff equal to $-(S(T_1) - K\exp(-r(T_2 - T_1)))$, since we sold the call and it is now ending in the money. At that point, we short one stock for the price of $S(T_1)$. All the cash is put in the bank account and totals $K\exp(-r(T_2 - T_1))$. Indeed, we got $S(T_1)$ from shorting the stock and $-(S(T_1) - K\exp(-r(T_2 - T_1)))$, the (negative) payoff from the short dated call. So now, we are short a stock and have in our bank account $K\exp(-r(T_2 - T_1))$. We then wait until T_2, at which point our bank account has grown to the value K. If the long dated call ends in the money ($S(T_2) < K$), we exercise it and receive the stock for which we pay the price K and hence our short position in stock (initiated at time T_1) is closed and the bank account is back to zero. If the long dated call ends OTM ($S(T_2) \leq K$), we buy a stock in the market for a value less than K and hence close our short position in stock and still have money in the bank account left, since $S(T_2) \leq K$. In all the above cases we started with nothing and had some free money after we bought and sold the two calls. In none of the cases did we have a loss at the end; in some we even had some extra profit. This is a clear arbitrage and hence our starting assumption is false. We must have that $EC(K\exp(-r(T_2 - T_1)), T_1) - EC(K, T_2) \leq 0$.

In a two-price world the inequality becomes:

$$bidEC(K\exp(-r(T_2 - T_1)), T_1) \leq askEC(K, T_2). \tag{1.3}$$

Indeed, suppose again the contrary, namely $bidEC(K\exp(-r(T_2 - T_1)), T_1) - askEC(K, T_2) > 0$; then again we could buy the long dated call (at the ask) and sell the short dated call (at the bid) and still have after these transactions some positive amount of money left. We further could exactly follow the above procedure, with no possibility of losing and always ending strictly positively. Hence again an arbitrage and therefore the Inequality (1.3) must hold.

A similar argument can be made to derive the calendar spread inequality for a put. In the one-price world the prices of two European Put options on the same underlier, but with different maturities $0 < T_1 < T_2$, satisfy

$$EP(K\exp(-r(T_2 - T_1)), T_1) \leq EP(K, T_2),$$

where we assumed no dividends. In a two-price world the inequality becomes:

$$bidEP(K\exp(-r(T_2 - T_1)), T_1) \leq askEP(K, T_2).$$

1.8 Dividends

Holding a stock entitles the owner of it to receive dividend payments. Usually, the stock price decreases by approximately the dividend amount after it is paid out (the ex-dividend date). Since dividend payments hence influence the stock price, the pricing of an equity derivative instrument should take into account dividends as well, and for this one needs to make an estimation of future dividend payments. This is, however, not straightforward, since dividends are decided at firm level and will depend on the future performance and strategy of the company. Dividends are usually expressed in a currency amount per stock. As is the case for interest rates, a continuously compounded dividend rate or yield q can be estimated. Using such a continuously compound dividend rate often simplifies the calculations and we will mainly use this setting.

Further, many of the formulae we have encountered so far can easily be adapted for such a continuously compound dividend regime. The classical put–call parity (Equation (1.2)), for example, becomes

$$\exp(-qT)S(0) - \exp(-rT)K + EP(K, T) - EC(K, T) = 0.$$

The underlying argument of buying one stock and holding it until time T is replaced with initial buying only $\exp(-qT)$ amount of stock and continuously investing the proceeds of dividends in additional stock, which again leads to one unit of stock at time T.

2

Stochastic Processes and Financial Models

2.1 Risk-Neutral Valuation

In financial engineering there are essentially two different worlds. The so-called real world (sometimes also referred to as the historical or physical world or P-world) and the pricing world (sometimes also referred to as the risk-neutral world or Q-world). We recognize up-front that in markets every event has in principle both a probability or a likelihood of its occurrence and a price, i.e. the price of a security paying a dollar if the event occurs. If that price is paid at event resolution, the event price we speak of is then a forward price. These forward prices are non-negative, being claims to non-negative cash flows. The prices for disjoint or mutually exclusive events are additive by no-arbitrage. They sum to unity over a set of mutually exclusive and exhaustive events as the forward price of a dollar for certain is a dollar. Hence, prices of events behave like probabilities (pricing world) and the mathematics of probability applies to them, but they are not probabilities; they are forward prices. They can differ substantially from the probabilities of the events actually occuring (real world). A typical example is found in insurance, where the forward premium paid for life, fire or car insurance by a particular individual far exceeds the probability of the insured event. The potential personal damage caused by the event induces people in markets to pay more than the probability for coverage. Competitive pressures do not reduce prices to probability assessments as sufficiently large pools of identical risks may not be available, leaving sellers exposed to risk that must be compensated in the premium. In these two worlds, one looks differently to the stochastic behaviour of the assets under investigation; the prices of events seen as probabilities (pricing world) differ from the probabilities of the events happening in the real world.

Most of the time, the probability measuring how things happen in the real world is denoted with a P; therefore also this real world is often named the P-world. This probability measure is measuring how things actually happen in reality, and one often refers to it as the physical measure. Therefore, one is typically estimating distributions in this world on the basis of historically observed real data of the underlying asset, like daily log-returns of a stock. In contrast, the risk-neutral world is the pricing world, created by financial markets trading event risk and modelled by financial engineers. Most of the time, the probability measure of how things happen in the pricing world is denoted with a Q and one refers to it as the pricing measure. One can prove that in absence of arbitrage, under this measure, traded assets all behave in a "risk-neutral" way, meaning that their expected return is equal to the return of the risk-free account. For example, for a stock with dividend yield q, we have that

$$E_Q[S_t] = \exp((r-q)t)S_0, \tag{2.1}$$

since the price of the forward stock net of intermediate dividends must equal the cost of buying it in the spot market on borrowed money and repaying the loan. One then has a rate of return q from the dividends and, since r is the rate of return on the risk-free account, the rate of return on a stock needs to be $r-q$, otherwise there would be the well-known spot forward arbitrage opportunity. The pricing world or Q-world is hence also often referred to as the risk-neutral world.

The reason for the existence of the particular risk-neutral world is the trading of risks in markets, thus leading to the existence of forward prices. Fundamental theory actually shows that under the no-arbitrage assumption (and some additional technical assumptions), the price of a derivative is given by the discounted expected payoff of the derivative, with expectations taken under the risk-neutral (Q) measure. One hence has that, for a European option maturing at time T, its current fair price is given by

$$\text{price} = \exp(-rT)E_Q[\text{payoff}].$$

Furthermore, the price of any traded event A resolved at T is

$$\text{price}(A) = \exp(-rT)Q(A),$$

with $Q(A)$ the probability under Q of A occurring. This is particularly useful, since calculating expected values is typically a doable operation in probability theory, and moreover, it is particularly suitable for a Monte Carlo setting.

It is not always easy to determine what the appropriate Q is and whether there are one or more measures under which traded assets behave on average like the risk-free account. The existence of a Q measure is (modulo some technical conditions) equivalent, with both the model and the market being free of arbitrage. The uniqueness of such a Q is then (modulo some technical conditions) equivalent with the market being complete, meaning that under frictionless circumstances each derivative can be perfectly hedged dynamically with positions in the underlying asset and the risk-free account. Nonetheless, the point of price discovery in markets is that, when traded, it is markets that determine these prices. It is their raison d'être. With well-functioning markets the Q is given uniquely by the market, and determining whether there are other measures under which assets are also behaving risk-neutrally is a purely theoretical exercise. In addition, one could argue that P is actually more a personal view, and therefore the uniqueness of P is also an issue of discussion.

The Black–Scholes model, introduced in the next section, is an example of an arbitrage-free and complete model. However, all other more advanced models are not complete, and there can be more (often infinitely many) measures under which the underlying asset behaves risk-neutrally. It is important, however, that there is one such special Q under which all the derivatives are priced, i.e. the price of any derivative on the underlying asset is given by the discounted expected payoff of the derivative under Q. This is therefore why one refers to such a measure as the pricing measure. This pricing measure is, as mentioned, given/determined by the market.

In many applications the actual P is irrelevant; one tries to estimate directly the special pricing measure Q via a so-called calibration procedure. The information to estimate Q comes from observable market prices of derivatives. Since these market quotes are discounted expectations under one particular Q of the related payoffs of the derivatives, one can try to estimate this underlying probability measure Q from these market quotes.

2.2 The Black–Scholes Model

This section provides an overview of the most basic and well-known continuous-time, continuous-variable stochastic model for stock prices. An understanding of this is the first step to the understanding of the pricing of options in a more advanced setting.

2.2.1 The Normal Distribution

The Normal distribution, Normal(μ, σ^2), is one of the most important distributions in many areas. It lives on the real line, has mean $\mu \in \mathbb{R}$ and variance $\sigma^2 > 0$ and its density function is given as

$$f_{Normal}(x; \mu, \sigma^2) = \frac{1}{\sqrt{2\pi\sigma^2}} \exp\left(-\frac{(x-\mu)^2}{2\sigma^2}\right). \tag{2.2}$$

In Figure 2.1, one sees the typical bell-shaped curve of a standard Normal (Normal$(0, 1)$) density.

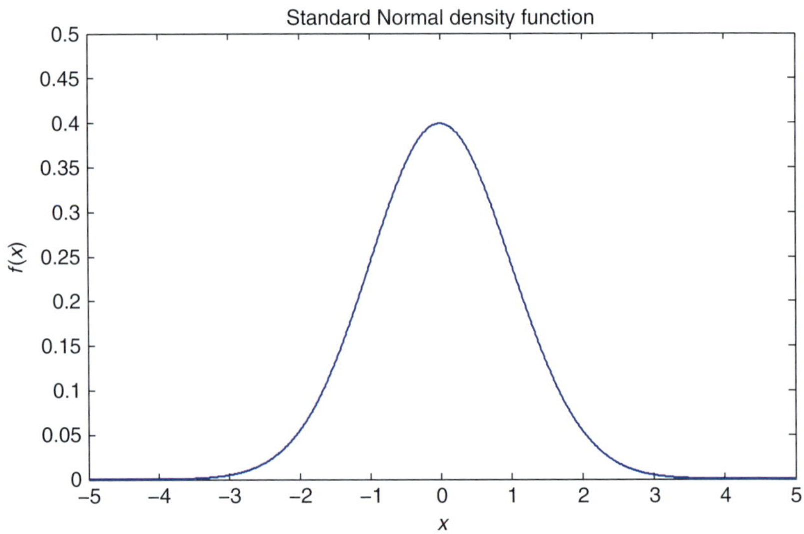

Figure 2.1 Density of a standard Normal distribution (Normal(0, 1))

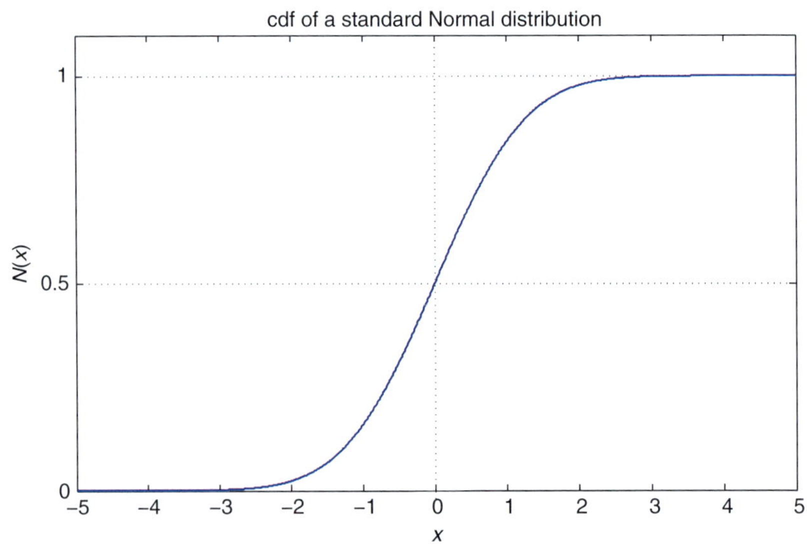

Figure 2.2 Cumulative distribution function of a standard Normal distribution (Normal(0, 1))

We will denote by

$$N(x) = \int_{-\infty}^{x} f_{Normal}(u; 0, 1)\mathrm{d}u \tag{2.3}$$

the cumulative distribution function (cdf) for a variable that is standard normally distributed (Normal(0, 1)). This special function is available in most mathematical software packages. The function is shown in Figure 2.2.

The Normal(μ, σ^2) distribution is symmetric around its mean, and always has a kurtosis of 3:

	Normal(μ, σ^2)
mean	μ
variance	σ^2
skewness	0
kurtosis	3

The Normal distribution arises from the Central Limit Theorem (CLT). Intuitively, it tells us that the suitable normalized sum of many independent random variables is approximately normally distributed. More precisely, we have that if X_i, $i = 1, 2, \ldots$, is a series of identically and independently distributed random variables with a finite second moment, we have that

$$P\left(\frac{\sum_{i=1}^{n} X_i - nE[X]}{\sqrt{n\text{Var}[X]}} \leq x\right) \to \text{N}(x), \text{ if } n \to \infty,$$

where $E[X]$ and $\text{Var}[X] = E[(X - E[X])^2]$ denote respectively the mean and the variance of the random variables X_i, $i = 1, 2, \ldots$. This result explains the ubiquity of the Normal distribution in a static context.

2.2.2 Brownian Motion

The big brother of the Normal distribution is Brownian motion. Brownian motion is the dynamic counterpart – where we work with evolution over time – of the static counterpart, the Normal distribution. The history of Brownian motion dates back to 1828, when the Scottish botanist Robert Brown observed pollen particles in suspension under a microscope and observed that they were in constant irregular motion. By doing the same with particles of dust, he was able to rule out that the motion was due to pollen being “alive”. In 1900, Louis Bachelier (1900) considered Brownian motion as a possible model for stock market prices. Bachelier’s model was his thesis. At that time the topic was not thought worthy of study. In 1923, Norbert Wiener defined and constructed Brownian motion rigorously for the first time. The resulting stochastic process is often called the Wiener process in his honour. It was with the work of Samuelson (1965) that Brownian motion reappeared as a modelling tool in finance.

Definition

A stochastic process $X = \{X_t, t \geq 0\}$ is a *standard Brownian motion* on some probability space $(\Omega, \mathcal{F}, P)$, if

1. $X_0 = 0$ a.s.
2. X has independent increments.

3. X has stationary increments.
4. $X_{t+s} - X_t$ is normally distributed with mean 0 and variance $s > 0$: $X_{t+s} - X_t \sim$ Normal$(0, s)$.

We shall henceforth denote standard Brownian motion by $W = \{W_t, t \geq 0\}$ (W for Wiener). Note that the second item in the definition implies that Brownian motion is a Markov process. Moreover, Brownian motion is the basic example of a Lévy process (see Section 2.4 and Schoutens (2003)). In Figure 2.3, we depict 10 paths of standard Brownian motions over the time interval [0, 1]. We refer to Section 3.3.1 for how to simulate such paths.

2.2.3 Geometric Brownian Motion

Now that we have Brownian motion W, we can introduce an important stochastic process for us, a relative of Brownian motion – *geometric Brownian motion*.

In the Black–Scholes model, one models the time evolution of a stock price $S = \{S_t, t \geq 0\}$ as follows. Consider how S will change in some small time interval from the present time t to a time $t + \Delta t$ in the near future. Writing ΔS_t for the change $S_{t+\Delta t} - S_t$, the relative return over this time step is then $\Delta S_t / S_t$. It is economically reasonable to expect this return to decompose into two components, a *systematic* part and a *random* part.

Let us first look at the systematic part. One assumes that the stock's expected return over a period is proportional to the length of the period considered. This means that, in a short interval of time $[t, t + \Delta t]$ of length Δt, the expected relative

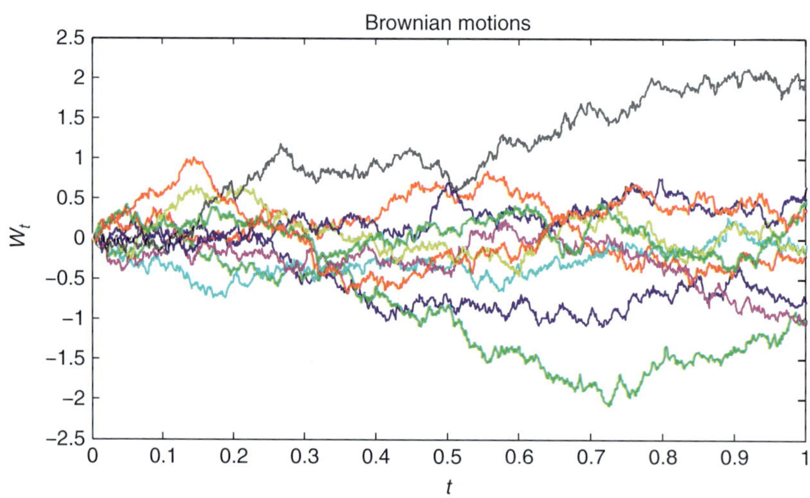

Figure 2.3 10 paths of a standard Brownian motion

return in S is proportional to the length of the time step, namely Δt, and is therefore given by $\mu \Delta t$, where μ is some parameter representing the mean rate of the return of the stock.

A stock price fluctuates stochastically, and a reasonable assumption is that the variance of the return over the interval of time $[t, t + \Delta t]$ is proportional to the length of the interval. Furthermore, taking into account the Central Limit Theorem and viewing the stock return as the sum of many random effects, it makes sense to model the random part of the return by a Normally distributed random variable with variance $\sigma^2 \Delta t$, with $\sigma > 0$ a parameter which describes how much effect this random part has – or how volatile the return is; σ is called the *volatility* of the stock. Putting this together and recognizing that $\sigma \Delta W_t = \sigma(W_{t+\Delta t} - W_t)$ has a normal distribution with the required variance $\sigma^2 \Delta t$, we could write

$$\Delta S_t = S_t(\mu \Delta t + \sigma \Delta W_t), \quad S_0 > 0.$$

In the limit, as $\Delta t \to 0$, we have the Stochastic Differential Equation (SDE)

$$\mathrm{d}S_t = S_t(\mu \mathrm{d}t + \sigma \mathrm{d}W_t), \quad S_0 > 0. \tag{2.4}$$

Using Itô's Lemma, one can prove that this stochastic differential equation has the unique solution

$$S_t = S_0 \exp\left(\left(\mu - \frac{\sigma^2}{2}\right) t + \sigma W_t\right).$$

This (exponential) functional of Brownian motion is called geometric Brownian motion. Note that

$$\log S_t - \log S_0 = \left(\mu - \frac{\sigma^2}{2}\right) t + \sigma W_t$$

has a Normal($t(\mu - \sigma^2/2), \sigma^2 t$) distribution. Thus S_t itself has a *log-normal* distribution. This geometric Brownian motion model, and the log-normal distribution which it entails, are the basis for the Black–Scholes model for stock-price dynamics in continuous time.

In Figure 2.4, one sees the realization of the geometric Brownian motion based on the sample path of the standard Brownian motion of Figure 2.3.

2.2.4 The Black–Scholes Option Pricing Model

In the early 1970s, Fischer Black, Myron Scholes and Robert C. Merton made a major breakthrough in the pricing of stock options by developing what has become known as the Black–Scholes model (Black and Scholes, 1973, and Merton, 1973). The model has had a huge influence on the way that traders price and hedge options.

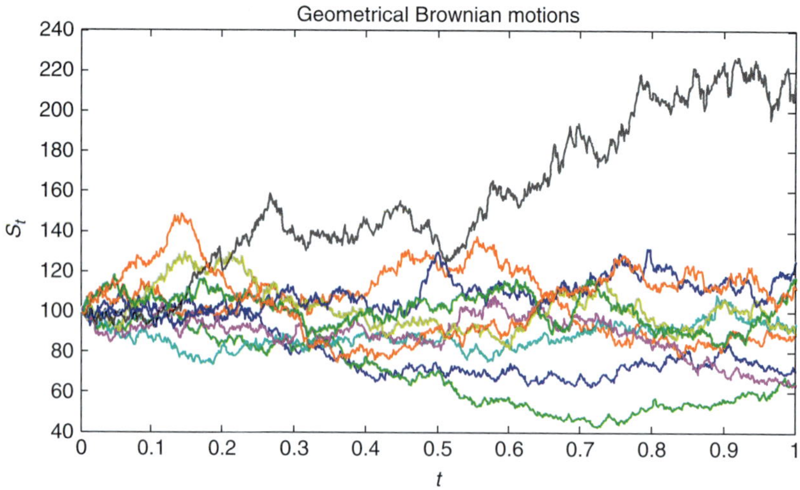

Figure 2.4 10 paths of a geometrical Brownian motion ($\mu = 0.05, \sigma = 0.40$)

In 1997, the importance of the model was recognized when Myron Scholes and Robert C. Merton were awarded the Nobel prize for economics. Sadly, Fischer Black died in 1995; otherwise he also would undoubtedly have been one of the recipients of this prize as well.

Next, we detail the Black–Scholes model and employ it for valuing European Call and Put options on a stock.

The Black–Scholes Market Model

Investors are allowed to trade continuously up to some fixed finite planning horizon T and the market has two basic assets.

The first asset is one without risk (the risk-free account). Its price process is given by $B = \{B_t = \exp(rt), 0 \leq t \leq T\}$. The second asset is a risky asset, usually referred to as stock, which pays a continuous dividend yield $q \geq 0$. The price process of this stock, $S = \{S_t, 0 \leq t \leq T\}$, is modelled by geometric Brownian motion:

$$S_t = S_0 \exp\left(\left(\mu - \frac{\sigma^2}{2}\right) t + \sigma W_t\right),$$

where $W = \{W_t, t \geq 0\}$ is standard Brownian motion.

Note that W_t has a Normal$(0, t)$ and that $S = \{S_t, t \geq 0\}$ satisfies the SDE (2.4). The parameter μ is reflecting the drift and σ models the volatility; μ and σ are assumed to be constant over time.

We call this market model the *Black–Scholes model*. It is a well-established result that the Black–Scholes model is a complete model; that is, every contingent claim can be replicated by a dynamic self-financing trading strategy.

The Risk-Neutral Setting

Since the Black–Scholes market model is complete there exists only one risk-neutral measure or pricing measure. One can show that under this risk-neutral measure Q the stock price is following again a geometric Brownian motion and has the same volatility parameter σ. The drift parameter μ, however, has to be equal to the continuously compounded risk-free rate r minus the dividend yield q to enforce 2.1:

$$S_t = S_0 \exp\left(\left(r - q - \frac{\sigma^2}{2}\right)t + \sigma W_t\right).$$

Equivalently, we can say that the risk-neutral stock price process $S = \{S_t, 0 \leq t \leq T\}$ is satisfying the SDE:

$$\mathrm{d}S_t = S_t((r-q)\mathrm{d}t + \sigma \mathrm{d}W_t), \quad S_0 > 0.$$

In a risk-neutral world the total return from the stock must be r; the dividends provide a return of q, the expected growth rate in the stock price is $r - q$.

Next, we will calculate European Call option prices under this model.

The Pricing of Options under the Black–Scholes Model

By the risk-neutral valuation principle the price V of a contingent claim with payoff function $G(\{S_u, 0 \leq u \leq T\})$ maturing at time T is given by

$$V = \exp(-Tr)E_Q[G(\{S_u, 0 \leq u \leq T\})]. \tag{2.5}$$

Furthermore, if the payoff function is only depending on the time T value of the stock, i.e. $G(\{S_u, 0 \leq u \leq T\}) = G(S_T)$, then the above formula can be rewritten as:

$$\begin{aligned} V &= \exp(-Tr)E_Q[G(S_T)] \\ &= \exp(-Tr)E_Q[G(S_0 \exp((r - q - \sigma^2/2)T + \sigma W_T))] \\ &= \exp(-Tr)\int_{-\infty}^{+\infty} G(S_0 \exp((r - q - \sigma^2/2)T + \sigma x)) f_{Normal}(x; 0, T)\mathrm{d}x, \end{aligned}$$

where expectations are taken under the pricing measure Q.

Explicit Formula for European Call and Put Options

In some cases it is possible to evaluate explicitly the expected value in the pricing formula (2.5).

Take, for example, a European call on the stock with price process S, strike K and maturity T (so $G(S_T) = (S_T - K)^+$). The Black–Scholes formula for the price

$EC(K, T)$ at time zero of this European Call option on the stock (with dividend yield q) is given by

$$EC(K, T) = \exp(-qT)S_0\mathrm{N}(d_1) - K\exp(-rT)\mathrm{N}(d_2),$$

where

$$d_1 = \frac{\log(S_0/K) + (r - q + \frac{\sigma^2}{2})T}{\sigma\sqrt{T}}, \tag{2.6}$$

$$d_2 = \frac{\log(S_0/K) + (r - q - \frac{\sigma^2}{2})T}{\sigma\sqrt{T}} = d_1 - \sigma\sqrt{T}, \tag{2.7}$$

and $\mathrm{N}(x)$ is the cumulative probability distribution function for a variable that is standard Normally distributed (Normal(0, 1)).

From this, one can also easily (via the put–call parity) obtain the price $EP(K, T)$ of the European Put option on the same stock with same strike K and same maturity T:

$$EP(K, T) = -\exp(-qT)S_0\mathrm{N}(-d_1) + K\exp(-rT)\mathrm{N}(-d_2).$$

For the call, the risk-neutral probability (Q) of finishing in the money corresponds with $\mathrm{N}(d_2)$. Similarly, the delta (i.e. the change in the value of the option compared with the change in the value of the underlying asset – see also Section 3.1.2) of the option corresponds with $\mathrm{N}(d_1)$.

2.2.5 Shortfalls of the Black–Scholes Model

Over the last decades the Black–Scholes model

$$S_t = S_0\exp((\mu - \sigma^2/2)t + \sigma W_t), \qquad t \geq 0,$$

where $\{W_t, t \geq 0\}$ is standard Brownian motion and $\sigma > 0$ is the usual volatility, turned out to be very popular. One should bear in mind, however, that this elegant theory hinges on several crucial assumptions. One assumes that there are no market frictions, like taxes and transaction costs or constraints on the stock holding, etc. Moreover, empirical evidence suggests that the classical Black–Scholes model does not describe the statistical properties of financial time series very well.

Summarizing, we could say that the Black–Scholes framework has several shortcomings which can have a serious impact on the modelling of financial assets and the corresponding pricing and hedging of financial derivatives. The most common critiques are:

- Log-returns under the Black–Scholes model are Normally distributed. However, it is observed from empirical data that log-returns typically do not behave according to a Normal distribution. Most of the time they show negative skewness and excess kurtosis.

- Related to the above observation on the log-returns, the Black–Scholes model cannot model realistically extreme events.
- Paths of the stock process under the Black–Scholes model are continuous and show no jumps. However, in reality, asset prices do jump, and the more pronounced jumps typically have the most impact on the derivative pricing.
- The volatility parameter (the only model parameter of relevance for the pricing of derivatives) is assumed to be constant. However, it has been observed that the volatilities or the parameters of uncertainty estimated (or more generally the environment) change stochastically over time.
- Stock prices may be seen as functions of a more meaningful concept of time related to the passage of economic activity as opposed to calendar time. Such economic measures of time, when they are increasing random processes, are purely discontinuous, making the continuity of stock prices a vacuous and fallacious proposition.

We refer to the literature and in particular to Schoutens (2003) for more details.

2.2.6 Implied Black–Scholes Volatility

A way to see that the classical Black–Scholes model cannot capture option prices in the market consistently is by looking at the implied volatilities coming from the market option prices. For every European Call option with strike K and time to maturity T, we calculate the only (free) parameter involved, the volatility $\sigma = \sigma(K, T)$, such that the theoretical option price (under the Black–Scholes model) matches the empirical one. This $\sigma = \sigma(K, T)$ is called the *implied volatility* of the option. Implied volatility is a timely measure – it reflects the market's perception at the time it is measured (and is not backward-looking based on historical data).

There is no closed form formula to extract the implied volatility out of the call option price. We have to rely on numerical methods. One method to find numerically implied volatilities is the classical Newton–Raphson iteration procedure. Denote by $EC_{BS}(\sigma)$ the price of the relevant call option as a function of volatility under the Black–Scholes model. If EC_{Market} is the market price of this option, we need to solve the equation

$$EC_{Market} = EC_{BS}(\sigma) \tag{2.8}$$

for σ. We start with some initial value we propose for σ; we denote this starting value with σ_0. It turns out that a σ_0 around 0.20 performs very well for most common stocks and indices. In general, if we denote by σ_n the value obtained after n iteration steps, the next value σ_{n+1} is given by

$$\sigma_{n+1} = \sigma_n - \frac{EC_{BS}(\sigma_n) - EC_{Market}}{EC'_{BS}(\sigma_n)},$$

where the function in the denominator, EC'_{BS}, refers to the differential with respect to σ of the call price function. This quantity is also referred to as the vega. For the European Call option (under Black–Scholes) we have:

$$EC'_{BS}(\sigma_n) = S_0\sqrt{T}\mathrm{N}(d_1) = S_0\sqrt{T}\mathrm{N}\left(\frac{\log(S_0/K) + (r - q + \frac{\sigma_n^2}{2})T}{\sigma_n\sqrt{T}}\right),$$

where S_0 is the current stock price, d_1 as in Equation (2.6) and $\mathrm{N}(x)$ is the cumulative probability distribution of a Normal(0, 1) random variable as in Equation (2.3).

In Figures 2.5 and 2.6 one sees the so-called volatility surface of the S&P 500 on 12 December 2014 and 17 July 2015, respectively (see also Section 3.2.1) based on closing mid-prices. Under the Black–Scholes model, all σs should be the same; clearly we observe that there is a huge variation in this volatility parameter both in strike as in time to maturity. One says often there is a volatility smile or skew effect. Again, this points to the fact that the Black–Scholes model is not appropriate and the traders already count this deficiency into their prices. Note, further, that both surfaces also differ quite significantly. Volatility, or in other words the parameters of uncertainty estimated (or more generally the environment), change stochastically over time.

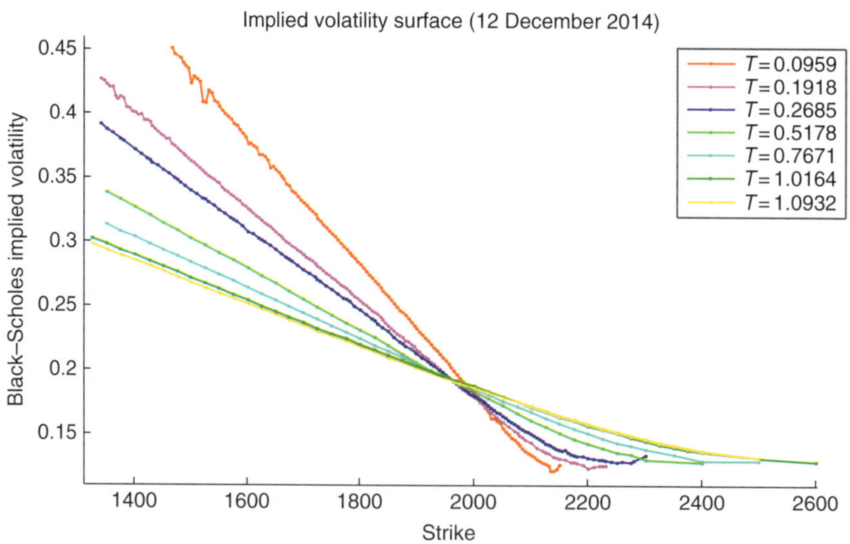

Figure 2.5 Implied volatility surface on 12 December 2014

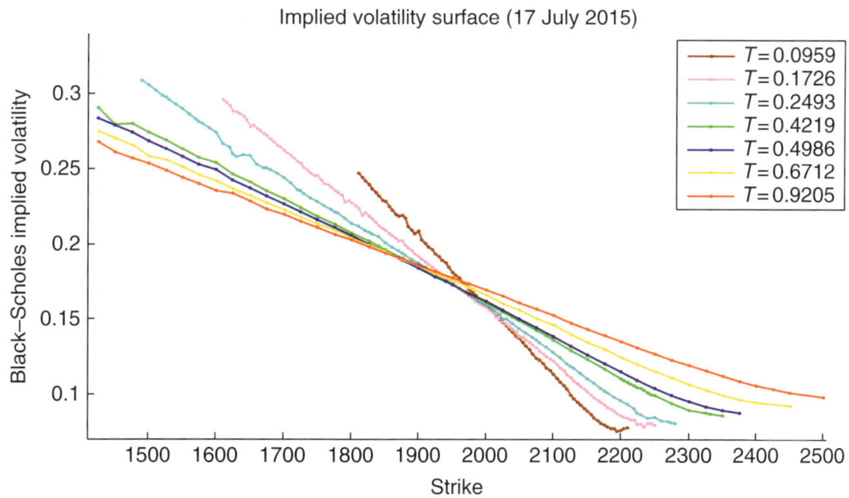

Figure 2.6 Implied volatility surface on 17 July 2015

2.3 Characteristic Functions

In this section we give more details about characteristic functions: essential mathematical objects in financial engineering.

Characteristic functions provide us a way to describe the dynamics/stochastics of some popular advanced models unambiguously. Moreover, out of them one can obtain many interesting properties of the underlying distribution, such as moments. Finally, and maybe for us most importantly, they serve as main input for vanilla pricing algorithms which will be constructed in Section 3.1.1.

The characteristic function ϕ of a distribution, or equivalently of a random variable X, is the Fourier–Stieltjes transform of the distribution function $F(x) = P(X \leq x)$:

$$\phi_X(u) = E[\exp(\mathrm{i}uX)] = \int_{-\infty}^{+\infty} \exp(\mathrm{i}ux)\mathrm{d}F(x),$$

where i is the imaginary number ($\mathrm{i}^2 = -1$).

If the random variable has a continuous distribution with a density function, say $f_X(x)$, then one can write:

$$\phi_X(u) = E[\exp(\mathrm{i}uX)] = \int_{-\infty}^{+\infty} \exp(\mathrm{i}ux) f_X(x)\mathrm{d}x.$$

Example 2.1 *The Normal distribution Normal*(μ, σ^2)*, with mean* μ *and variance* σ^2 *lives on* $(-\infty, +\infty)$ *and has a density function given by Equation (2.2). Its characteristic function is therefore given by*

$$\phi_X(u) = \int_{-\infty}^{+\infty} \exp(\mathrm{i}ux)\frac{1}{\sqrt{2\pi\sigma^2}} \exp\left(-\frac{(x-\mu)^2}{2\sigma^2}\right) dx$$
$$= \exp(\mathrm{i}u\mu)\exp\left(-\frac{1}{2}\sigma^2 u^2\right).$$

Let us derive the above formula for the Normal$(0,1)$ case. We will comment later on how to derive the general situation. We have

$$\begin{aligned}
\phi_X(u) &= \int_{-\infty}^{+\infty} \exp(\mathrm{i}ux)\frac{1}{\sqrt{2\pi}} \exp\left(-\frac{x^2}{2}\right) dx \\
&= \frac{1}{\sqrt{2\pi}}\int_{-\infty}^{+\infty} \exp\left(-\frac{x^2}{2} + \mathrm{i}ux\right) dx \\
&= \frac{1}{\sqrt{2\pi}}\int_{-\infty}^{+\infty} \exp\left(-\left((x/\sqrt{2})^2 - \mathrm{i}ux + (\mathrm{i}u/\sqrt{2})^2 - (\mathrm{i}u/\sqrt{2})^2\right)\right) dx \\
&= \frac{\exp(-u^2/2)}{\sqrt{2\pi}}\int_{-\infty}^{+\infty} \exp\left(-\left((x/\sqrt{2} - \mathrm{i}u/\sqrt{2})^2\right)\right) dx \\
&= \frac{\exp(-u^2/2)}{\sqrt{2\pi}}\sqrt{2}\int_{-\infty}^{+\infty} \exp(-z^2)dz \\
&= \frac{\exp(-u^2/2)}{\sqrt{2\pi}}\sqrt{2}\sqrt{\pi} \\
&= \exp(-u^2/2),
\end{aligned}$$

where we used in the penultimate line the formula for the so-called Gauss integral: $\int_{-\infty}^{+\infty}\exp(-z^2)dz = \sqrt{\pi}$.

For any distribution, the characteristic function always exists, is continuous and determines the distribution function uniquely. Moreover, we have

$$\phi_X(0) = E[\exp(\mathrm{i}0X)] = E[1] = 1.$$

Furthermore, we have

$$\phi_{-X}(u) = E[\exp(\mathrm{i}u(-X))] = E[\exp(\mathrm{i}(-u)X)] = \phi_X(-u). \tag{2.9}$$

More generally, we have that for any real constant a,

$$\phi_{aX}(u) = E[\exp(\mathrm{i}u(aX))] = E[\exp(\mathrm{i}(au)X)] = \phi_X(au). \tag{2.10}$$

From ϕ_X one can easily derive the moments of X. Indeed, if $E[|X|^k] < \infty$, then

$$E[X^k] = \mathrm{i}^{-k}\,\frac{\mathrm{d}}{\mathrm{d}u^k}\phi_X(u)\bigg|_{u=0}.$$

A very convenient property is the fact that the sum of two independent random variables, or equivalently the convolution of two distributions, translate into the product of the corresponding characteristic functions. More precisely, if X and Y are two independent random variables with characteristic functions ϕ_X and ϕ_Y, respectively, then the characteristic function of $Z = X + Y$ is given by

$$\phi_Z(u) = \phi_X(u)\phi_Y(u).$$

Finally, we note that the degenerate random variable $X \equiv \mu$, i.e. the random variable that takes with probability one the constant value μ (so there is no randomness), has characteristic function

$$E[\exp(\mathrm{i}uX)] = \exp(\mathrm{i}u\mu).$$

Example 2.2 *Using the above properties, we can easily deduce the characteristic function of a random variable Z with the Normal(μ, σ^2) distribution. Indeed, write $Z = \mu + \sigma X$ with X standard Normal. Then*

$$\begin{aligned}\phi_Z(u) &= \phi_{\mu+\sigma X}(u)\\ &= \phi_\mu(u)\phi_{\sigma X}(u)\\ &= \phi_\mu(u)\phi_X(\sigma u)\\ &= \exp(\mathrm{i}u\mu)\exp(-(\sigma u)^2/2).\end{aligned}$$

2.4 Lévy Processes

Suppose $\phi(x)$ is the characteristic function of a random variable X and that for every positive integer n, $\phi(x)$ is also the nth power of a characteristic function; we say that the underlying distribution is *infinitely divisible*. This means that we then can actually divide X into n identically and independently distributed parts, and this for any n we wish; the characteristic function of such a part has $\phi(x)^{1/n}$ as a characteristic function, which is by definition a proper characteristic function.

One then defines, for every such infinitely divisible distribution, a stochastic process, $X = \{X_t, t \geq 0\}$, called a *Lévy process*, which starts at zero and has independent and stationary increments such that the distribution of an increment $X_{t+s} - X_s$ over the time period $[s, s+t]$,$s, t \geq 0$ has $\phi(x)^t$ as its characteristic function.

The function $\log\phi(x)$ is often called the *characteristic exponent*, and one can prove that it satisfies the following Lévy–Khintchine formula (see Schoutens, 2003)

$$\log\phi(x) = \mathrm{i}\gamma u - \frac{1}{2}\sigma^2u^2 + \int_{-\infty}^{+\infty}\left(\exp(\mathrm{i}ux) - 1 - \mathrm{i}ux_{|x|<1}\right)\nu(\mathrm{d}x),$$

with $\gamma \in \mathbb{R}$, $\sigma > 0$ and $\nu(\mathrm{d}x)$ a measure on the real line. We say that our infinitely divisible distribution has a triplet of Lévy characteristics (or *Lévy triplet* for short),

$[\gamma, \sigma^2, \nu(\mathrm{d}x)]$. The measure ν is called the Lévy measure of X. If the Lévy measure is of the form $\nu(\mathrm{d}x) = u(x)dx$, we call $u(x)$ the Lévy density. The Lévy density has the same mathematical requirements as a probability density, except that it does not need to be integrable and must have zero mass at the origin. The Lévy measure $\nu(\mathrm{d}x)$ dictates how the jumps occur. Jumps of sizes in the set A occur according to a Poisson process with intensity parameter $\int_A \nu(\mathrm{d}x)$, or, in other words, they have inter-arrival times following an exponential distribution with mean equal to one over the intensity parameter.

From the Lévy–Khintchine formula, one can see that, in general, a Lévy process consists of three independent parts: a linear deterministic part, a Brownian part and a pure jump part.

2.5 The VG Model

In the previous chapter we saw that the Black–Scholes model has many imperfections. In this section we explore an alternative model: the so-called Variance Gamma (VG) model, which is driven by a particular Lévy process, namely the VG process. A whole battery of other (infinitely divisible) distributions and related Lévy processes, like CGMY, Meixner, NIG and GH process, can be investigated as well; we refer to Schoutens (2003) for an introduction, details and further applications of Lévy processes.

2.5.1 The VG Distribution

The Gamma distribution will be an essential building block of the construction of the Variance Gamma (VG) distribution, which we will deploy a lot throughout this book. We start with the definition and some properties of the Gamma distribution.

The Gamma Distribution

The Gamma distribution is a distribution that lives on the positive real numbers and depends on two parameters. More precisely, the density function of the Gamma distribution Gamma(a, b), with parameters $a > 0$ and $b > 0$, is given by

$$f_{Gamma}(x; a, b) = \frac{b^a}{\Gamma(a)} x^{a-1} \exp(-xb), \quad x > 0.$$

A special case is the Exponential distribution, which one obtains by taking $a = 1$. For different parameter values the density function is graphed in Figure 2.7.

The characteristic function is given by

$$\phi_{Gamma}(u; a, b) = (1 - \mathrm{i}u/b)^{-a}.$$

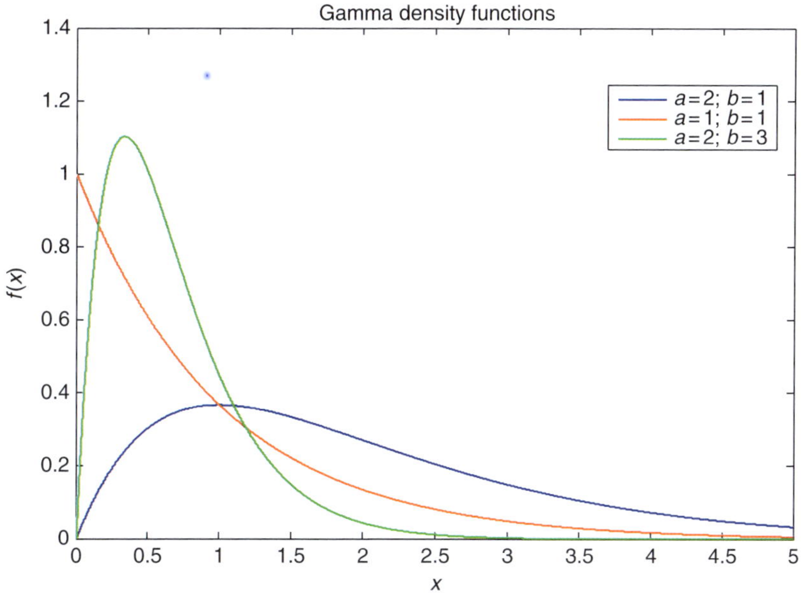

Figure 2.7 Gamma densities

The Gamma distribution is the basic example of a distribution with a so-called semi-heavy (right) tail. In general, we say that distributions or their density functions $f(x)$ have semi-heavy tails, if the tails of the density function behave as

$$\begin{aligned} f(x) &\sim C_-|x|^{\rho_-}\exp(-\eta_-|x|) \quad \text{as} \quad x \to -\infty \\ f(x) &\sim C_+|x|^{\rho_+}\exp(-\eta_+|x|) \quad \text{as} \quad x \to +\infty, \end{aligned}$$

for some $\rho_-, \rho_+ \in \mathbb{R}$ and $C_-, C_+, \eta_-, \eta_+ \geq 0$. Equivalently,

$$\begin{aligned} \log f(x) &\sim A_- \log|x| - \eta_-|x| \quad \text{as} \quad x \to -\infty \\ \log f(x) &\sim A_+ \log|x| - \eta_+|x| \quad \text{as} \quad x \to +\infty, \end{aligned}$$

for some $A_-, A_+ \in \mathbb{R}$ and $\eta_-, \eta_+ \geq 0$.

Note that the Normal distribution with mean μ and variance σ^2 exhibits a quadratic decay near infinity of the logarithm of its probability density function:

$$\log f_{\text{Normal}}(x; \mu, \sigma^2) = -\frac{(x-\mu)^2}{2\sigma^2} - \log\left(\sigma\sqrt{2\pi}\right) \sim -\frac{1}{2\sigma^2}x^2 \tag{2.11}$$

as $x \to \pm\infty$. This quadratic decay of the tails is one of the many criticisms of the Black–Scholes model.

The following properties of the Gamma(a, b) distribution can easily be derived from the characteristic function:

	Gamma(a, b)
mean	a/b
variance	a/b^2
skewness	$2a^{-1/2}$
kurtosis	$3(1 + 2a^{-1})$

Note also that we have the following scaling property: if X is Gamma(a, b), then for $c > 0$, cX is Gamma$(a, b/c)$.

The VG Distribution

The VG distribution goes back to work by Madan and co-authors: see Madan and Seneta (1987), Madan and Seneta (1990), Madan and Milne (1991), Madan et al. (1998); see also Fu et al. (2000).

The Variance Gamma VG(C, G, M) distribution on $(-\infty, +\infty)$ can be constructed as the difference of two gamma random variables. Suppose that X is a Gamma$(a = C, b = M)$ random variable and that Y is a Gamma$(a = C, b = G)$ random variable and that they are independent of each other. Then

$$X - Y \sim \mathrm{VG}(C, G, M).$$

To derive the characteristic function, we start by noting that

$$\phi_X(u) = (1 - \mathrm{i}u/M)^{-C} \text{ and } \phi_Y(u) = (1 - \mathrm{i}u/G)^{-C}.$$

By using the property (2.9), we have

$$\phi_{-Y}(u) = (1 + \mathrm{i}u/G)^{-C}.$$

Summing the two independent random variables X and $-Y$ and using the convolution property (2.10) gives

$$\phi_{X-Y}(u) = (1 - \mathrm{i}u/M)^{-C}(1 + \mathrm{i}u/G)^{-C} = \left(\frac{GM}{GM + (M - G)\mathrm{i}u + u^2}\right)^C.$$

The VG distribution is infinitely divisible and the Lévy measure of the Lévy triplet is given by

$$\nu(\mathrm{d}x) = C\frac{\exp(-Mx)}{x}\mathrm{d}x, \text{ if } x > 0 \tag{2.12}$$

$$= C\frac{\exp(Gx)}{|x|}\mathrm{d}x, \text{ if } x < 0. \tag{2.13}$$

Another way of introducing the Variance Gamma (VG) distribution is by mixing a Normal distribution with a Gamma random variate. The procedure is as follows. Take a random variate $G \sim$ Gamma$(a = 1/\nu, b = 1/\nu)$. Note that $E[G] = 1$. Then sample a random variate $X \sim$ Normal$(\theta G, \sigma^2 G)$; then X follows a Variance

Gamma distribution. The distribution of X is denoted VG(σ, ν, θ) and thus depends on three parameters:

- a real number θ (in the mean of the Normal distribution);
- a positive number σ (in the variance of the Normal distribution);
- a positive number ν (of the Gamma random variable G).

One can show, using basic probabilistic techniques, that under this parameter setting, the characteristic function of the VG(σ, ν, θ) law is given by

$$E[\exp(\mathrm{i}uX)] = \phi_{VG}(u; \sigma, \nu, \theta) = (1 - \mathrm{i}u\theta\nu + \sigma^2\nu u^2/2)^{-1/\nu}.$$

Using elementary calculus one can find the correspondence between the two parameter settings. On the one hand, we could go from the (σ, ν, θ) setting to the parametrization in terms of $C(arr)$, $G(eman)$ and $M(adan)$ using

$$\begin{aligned} C &= 1/\nu > 0 \\ G &= \left(\sqrt{\frac{\theta^2\nu^2}{4} + \frac{\sigma^2\nu}{2}} - \frac{\theta\nu}{2}\right)^{-1} > 0 \\ M &= \left(\sqrt{\frac{\theta^2\nu^2}{4} + \frac{\sigma^2\nu}{2}} + \frac{\theta\nu}{2}\right)^{-1} > 0. \end{aligned}$$

Going the other way around, one can use:

$$\begin{aligned} \nu &= 1/C \\ \sigma^2 &= 2C/(MG) \\ \theta &= C(G - M)/(MG). \end{aligned}$$

The VG density function is given by

$$f_{\mathrm{VG}}(x; C, G, M)(x) = \frac{(G\,M)^C}{\sqrt{\pi}\,\Gamma(C)} \exp\left(\frac{(G - M)x}{2}\right) \times \left(\frac{|x|}{G + M}\right)^{C-1/2} K_{C-1/2}\big((G + M)\,|x|/2\big),$$

where $\mathrm{K}_\nu(x)$ denotes the modified Bessel function of the third kind with index ν and $\Gamma(x)$ denotes the gamma function. As shown in Figure 2.8, one can see that the distribution is very flexible.

Some distribution characteristics are summarized in Tables 2.1 and 2.2. When $\theta = 0$ the distribution is symmetric. Negative values of θ result in negative skewness; positive θs give positive skewness. The parameter ν primarily controls the kurtosis. In terms of the (C, G, M)-parameters, $G = M$ gives the symmetric case,

Table 2.1 *VG distribution characteristics in the* (σ, ν, θ) *parametrization.*

	$\mathbf{VG}(\sigma, \nu, \theta)$	$\mathbf{VG}(\sigma, \nu, 0)$
mean	θ	0
variance	$\sigma^2 + \nu\theta^2$	σ^2
skewness	$\theta\nu(3\sigma^2 + 2\nu\theta^2)/(\sigma^2 + \nu\theta^2)^{3/2}$	0
kurtosis	$3(1 + 2\nu - \nu\sigma^4(\sigma^2 + \nu\theta^2)^{-2})$	$3(1 + \nu)$

Table 2.2 *VG distribution characteristics in the* (C, G, M) *parametrization.*

	$\mathbf{VG}(C, G, M)$	$\mathbf{VG}(C, G, G)$
mean	$C(G - M)/(MG)$	0
variance	$C(G^2 + M^2)/(MG)^2$	$2CG^{-2}$
skewness	$2C^{-1/2}(G^3 - M^3)/(G^2 + M^2)^{3/2}$	0
kurtosis	$3(1 + 2C^{-1}(G^4 + M^4)/(M^2 + G^2)^2)$	$3(1 + C^{-1})$

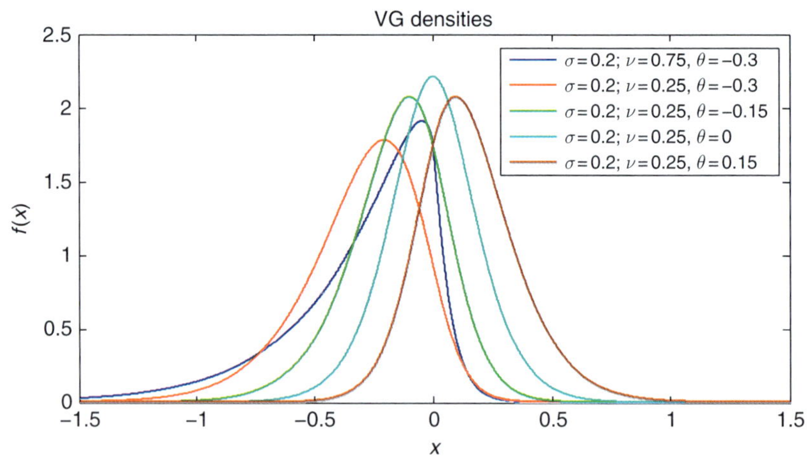

Figure 2.8 VG densities

$G < M$ results in negative skewness and $G > M$ give rise to positive skewness. The parameter C controls primarily the kurtosis.

A very special case is the VG distribution with parameters $\nu = 1$ and $\theta = 0$, or equivalently $C = 1$ and $G = M$. We then obtain the double-sided exponential distribution with density function as graphed in Figure 2.9:

$$f(x) = \frac{G}{2}\exp(-G|x|) = \frac{1}{\sqrt{2}\sigma}\exp(-\sqrt{2}|x|/\sigma).$$

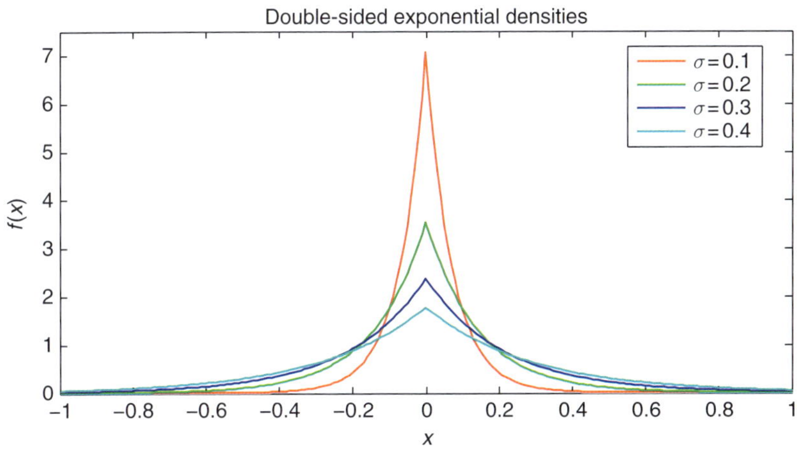

Figure 2.9 Density function of the double-sided exponential distribution

In, for example, Schoutens (2003), it is shown that, in contrast to a Normal distribution, a VG distribution can be fitted quite well on many historical data sets of (daily) log-returns. In Corcuera et al. (2009), the concept of Black–Scholes implied volatility of Section 2.2.6 is extended using, among others, the above double-sided exponential distribution. In 2002, M was looking at its VG model and was saying "Mmmh, something is missing". He called Y to explain "Why" he needed another parameter. Then G came along and said "Gee", this would be nice. Then, Y extended the model, things were explained to C, who immediately said "I see" and the VG model was extended to the CGMY model in Carr et al. (2002).

2.5.2 The VG Process

Recall the definition of a standard Brownian motion $W = \{W_t, t \geq 0\}$:

- W starts at zero: $W_0 = 0$.
- W has independent increments: the distribution of increments over non-overlapping time intervals are stochastically independent.
- W has stationary increments: the distribution of an increment over a time-interval depends only on the length of the interval, not on the exact location.
- $W_{s+t} - W_t \sim \text{Normal}(0, s)$: increments are Normally distributed.

One can define in a similar way a stochastic process based on the VG distribution. (For mathematical details and other examples see Schoutens (2003).) A stochastic process $X = \{X_t, t \geq 0\}$ is a Variance-Gamma process with parameters C, G, M if:

- X starts at zero: $X_0 = 0$.
- X has independent increments.
- X has stationary increments.
- Furthermore we have that $X_{s+t} - X_t \sim \text{VG}(Cs, G, M)$, i.e. increments are VG distributed.

It turns out (see Schoutens (2003)) that a VG process is a pure jump process. Sample paths have no diffusion component, in contrast with a Brownian motion, as can be seen in Figure 2.10. For simulation details we refer to Section 3.3.3.

2.5.3 The VG Stock Price Model

Instead of modelling the stock price process as an exponential of a Brownian motion (with drift):

$$S_t = S_0 \exp((\mu - \sigma^2/2)t + \sigma W_t), \qquad S_0 > 0,$$

we now model S as the exponential of a VG process $X = \{X_t, t \geq 0\}$:

$$S_t = S_0 \exp(X_t), \qquad S_0 > 0.$$

In that way, log-returns are no longer Normally distributed but follow the more flexible VG distribution:

$$\log S_{t+1} - \log S_t = X_{t+1} - X_t \sim VG(C, G, M), \qquad C, G, M > 0.$$

Note that under Black–Scholes we had:

$$\log S_{t+1} - \log S_t \sim \text{Normal}\left(\mu - \frac{\sigma^2}{2}, \sigma^2\right).$$

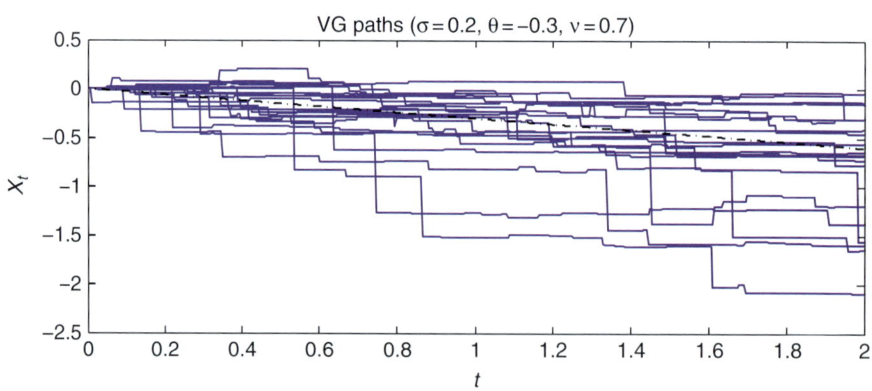

Figure 2.10 20 VG sample paths ($\sigma = 0.2$, $\nu = 0.75$ and $\theta = -0.3$)

Under a Black–Scholes framework, moving from a historical world to a risk-neutral one is easy: one replaces the drift μ with the interest rate r (minus the dividend yield q):

$$S_t = S_0 \exp((r - q - \sigma^2/2)t + \sigma W_t), t \geq 0.$$

In contrast with the Black–Scholes world, for the VG model (and in general for all more advanced models) there is no unique transformation. Actually, there are infinitely many possible measure changes. One particular easy transformation is the mean-correcting measure change, where the VG process is shifted in order to satisfy 2.1:

$$S_t = S_0 \exp((r - q)t) \frac{\exp(X_t)}{E[\exp(X_t)]} = S_0 \exp((r - q + \omega)t + X_t), t \geq 0,$$

where

$$\omega = -\log(E[\exp(X_t)])/t = -\log(E[\exp(X_1)]) = \nu^{-1} \log\left(1 - \frac{1}{2}\sigma^2\nu - \theta\nu\right).$$

Note that, most of the time, we will work immediately under such a risk-neutral setting (after calibrating the model to market data).

2.6 Sato Processes and Sato Models

2.6.1 Sato Processes

Sato processes are closely linked to the class of self-decomposable laws, which are a subclass of the infinitely divisible distributions. A detailed overview of self-decomposability and Sato processes can be found in Sato (2000). We recall here the main definitions and properties.

The distribution of a random variable X is self-decomposability if, for any constant c, $0 < c < 1$, X has the same probability law as the sum of a scaled version of itself and an independent random variable X_c:

$$X =^{law} cX + X_c.$$

Additive processes are stochastic processes with inhomogeneous and independent increments. In the special case where the increments are homogeneous we are dealing with Lévy processes.

A stochastic process $\{X_s, s \geq 0\}$ is called a self-similar process if, for any $\alpha > 0$, there exists a $\beta(\alpha) > 0$ such that for all $s > 0$

$$X_{\alpha s} =^{law} \beta(\alpha) X_s. \tag{2.14}$$

Lamperti (1962) proved that $\beta(\alpha)$ can be expressed as $\beta(\alpha) = \alpha^\gamma$, with $\gamma \geq 0$. The process is said to be self-similar of index γ.

Sato (2000) showed that a distribution is self-decomposable if and only if it is the distribution at unit time of a self-similar additive process.

Hence one can construct, from any self-decomposable distribution, a self-similar additive process; we call such a self-similar additive process a Sato process. Indeed, if X is self-decomposable, then there must be a self-similar additive process, $\{X_s, s \geq 0\}$ say, such that $X_1 =^{law} X$. But then we have, by using Equation (2.14), with $s = 1$ and $\alpha = t \geq 0$, that the probability law of the Sato process at time t is obtained by scaling the self-decomposable law of X:

$$X_t =^{law} t^{\gamma} X, \tag{2.15}$$

where γ is the self-similarity index.

2.6.2 The VG Sato Model

As an example of a Sato process, we introduce the VG Sato process. One can show that the VG is self-decomposable and a VG Sato process $Y = \{Y_t, t \geq 0\}$ can hence be constructed. By the scaling property (2.15), the law of Y_t is that of $t^{\gamma} X$, with X our mother VG distribution and hence

$$\begin{aligned}
\phi_{Y_t}(u, t; \sigma, \theta, \nu, \gamma) &= E[\exp(\mathrm{i}uY_t)] \\
&= E[\exp(\mathrm{i}ut^{\gamma} X)] \\
&= \phi_X(ut^{\gamma}; \sigma, \theta, \nu) \\
&= \left(1 - \mathrm{i}ut^{\gamma}\theta\nu + \sigma^2\nu u^2 t^{2\gamma}/2\right)^{-1/\nu}.
\end{aligned}$$

Note that hence

$$Y_t \sim \mathrm{VG}(\sigma t^{\gamma}, \nu, t^{\gamma}\theta).$$

We now model the risk-neutral stock price process as an exponential of the Sato process:

$$S_t = S_0 \exp((r-q)t)\frac{\exp(Y_t)}{E[\exp(Y_t)]} = S_0 \exp((r-q)t)\frac{\exp(Y_t)}{\phi_{Y_t}(-\mathrm{i}, t; \sigma, \theta, \nu, \gamma)}, \quad t \geq 0.$$

A simple calculation shows that

$$\phi_{Y_t}(-\mathrm{i}, t; \sigma, \theta, \nu, \gamma) = \left(1 - t^{\gamma}\theta\nu - \sigma^2\nu t^{2\gamma}/2\right)^{-1/\nu}.$$

2.7 Tree Models

Binomial approximations to the log-normal model have been extensively employed in option pricing from their introduction in Cox, Ross and Rubinstein (1979).

Extensions to Trinomial models have been considered, for example, in Boyle (1988), Kamrad and Ritchken (1991). In this section we review some basic (binomial and multinomial) tree models and consider approximation to VG models.

2.7.1 The Binomial Tree

We start with setting up the most simple tree setting: a one-step binomial tree. Assume the stock price is currently at S_0. We assume an interest rate r, and for the simplicity of the exposition we assume no dividends. The one-step binomial tree model allows the stock price after a time-step of $\Delta t > 0$ to move to only two possible values, namely uS_0 and dS_0, representing respectively the up-state and down-state.

In order to exclude arbitrage, we assume $0 \leq d < \exp(r\Delta t) < u$. The factors u and d are often chosen in terms of a given (yearly) volatility estimate σ. A common choice is the so-called Cox–Ross–Rubinstein setting where

$$u = \exp(\sigma\sqrt{\Delta t}) \text{ and } d = \exp(-\sigma\sqrt{\Delta t}).$$

Consider now a derivative paying out $f_u \geq 0$ in the up-state and $f_d \geq 0$ in the down-state as in Figure 2.11.

The current price f of this derivative can be obtained as the discounted expected payoff under the risk-neutral measure:

$$f = \exp(-r\Delta t)\,(pf_u + (1-p)f_d)\,,$$

where the risk-neutral probability to move to the up-state is given by

$$p = \frac{\exp(r\Delta t) - d}{u - d}.$$

Note that due to our no-arbitrage conditions, this is a proper probability: $0 < p < 1$.

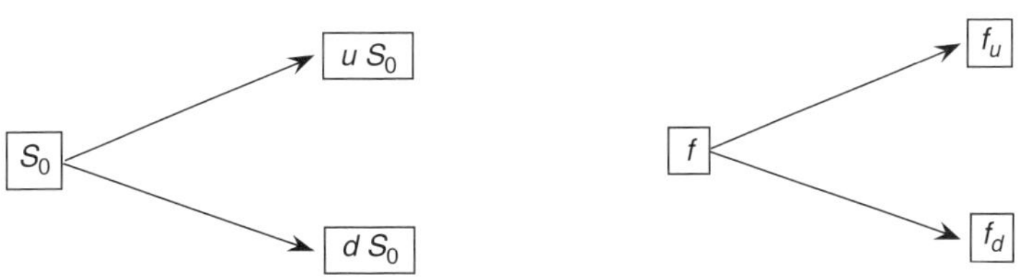

Figure 2.11 Binomial tree

Figure 2.12 Trinomial tree

2.7.2 The Trinomial Tree

In a trinomial tree, there is an additional middle-state and the stock can now jump after a time-step of $\Delta t > 0$ to three possible values, namely uS_0, mS_0 and dS_0, representing respectively the up-state, middle-state and down-state.

A common choice (see Derman et al., 1996) for the factors u, m and d in terms of a given (yearly) volatility estimate σ is

$$\begin{aligned} u &= \exp((r - \sigma^2/2)\Delta t + \sigma\sqrt{3\Delta t}) \\ m &= \exp((r - \sigma^2/2)\Delta t) \\ d &= \exp((r - \sigma^2/2)\Delta t - \sigma\sqrt{3\Delta t}). \end{aligned} \tag{2.16}$$

Consider now a derivative paying out f_u in the up-state, f_m in the middle-state and f_d in the down-state as in Figure 2.12.

A popular choice of corresponding probabilities is:

$$p_u = 1/6, \quad p_m = 2/3 \quad \text{and} \quad p_d = 1/6, \tag{2.17}$$

where p_u, p_m and p_d are respectively the probabilities to move to the up-, middle- or down-state. One can show that these probabilities provide us with an almost (but not exact) risk-neutral setting (compare with Equation (2.1)). Therefore, one often refers to these probabilities as the risk-neutral ones and the current price f of this derivative can be obtained, or better approximated, as the discounted expected payoff under the "risk-neutral" measure.

$$f \approx \exp(-r\Delta t)\,(p_u f_u + p_m f_m + p_d f_d)\,.$$

2.7.3 Multi-Step Trees

One-step trees can be combined into multi-step trees. For example, let us set up a two-step trinomial tree. Now, at each time-step Δt one can move from each state to one of three possible states. For the sake of ease, we assume these moves are always by the factors u, m or d as in Equation (2.16).

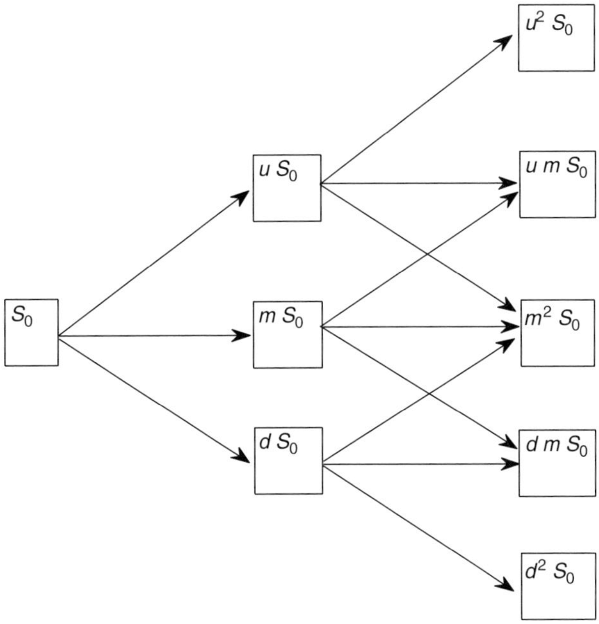

Figure 2.13 Two-step trinomial tree (stock values)

This leads to a (self-combining) two-step trinomial tree as in Figure 2.13. We further assume that in each step p_u, p_m and p_d are as in Equation (2.17) and respectively are the probabilities to move to the corresponding up-, middle- or down-state.

Now, one can calculate (or approximate) the price, as illustrated in Figure 2.14. We first calculate the prices in the intermediate states. The prices in the three possible intermediate states are given by

$$f_u \approx \exp(-r\,\Delta t)\,(p_u\,f_{uu} + p_m\,f_{um} + p_d\,f_{mm})$$
$$f_m \approx \exp(-r\,\Delta t)\,(p_u\,f_{um} + p_m\,f_{mm} + p_d\,f_{md})$$
$$f_d \approx \exp(-r\,\Delta t)\,(p_u\,f_{mm} + p_m\,f_{md} + p_d\,f_{dd})\,.$$

After these are calculated we can move one step backward to the initial state and calculate the initial price as

$$f \approx \exp(-r\,\Delta t)\,(p_u\,f_u + p_m\,f_m + p_d\,f_m)\,.$$

2.7.4 Multinomial Trees

We can also study multinomial trees, often used as approximations to one-dimensional processes that have exposure to many moves in an instant. For additional information see Yamada and Primbs (2004), Maller et al. (2006). We

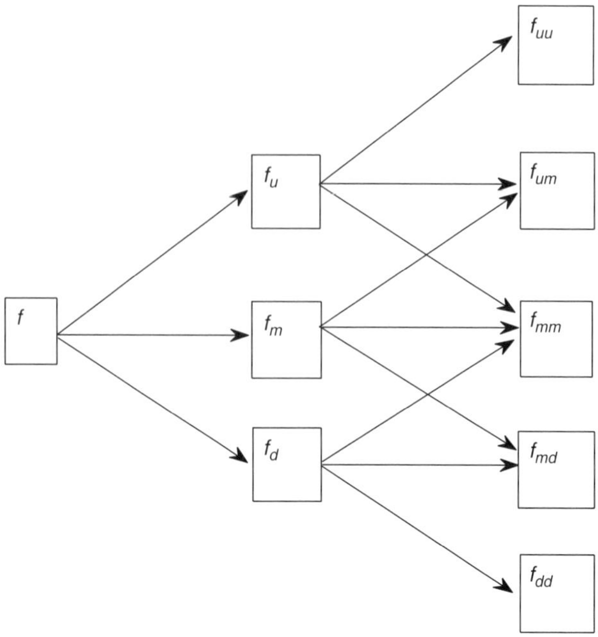

Figure 2.14 Two-step trinomial tree (derivative values)

proceed here to set up multinomial trees and show how to construct multinomial approximations to Lévy processes and illustrate this by considering the VG case. To do this, we will take the local motion to be multinomial. The goal is to construct multinomial approximations to a Lévy process at unit time by splitting unit time into N intervals of length $\Delta t = 1/N$.

Suppose we want to approximate a Lévy process $X = \{X_t, t \geq 0\}$ with characteristic function

$$\phi(u) = E\left[\exp(\mathrm{i}uX_1)\right].$$

Let $k(x)$ be the Lévy density and let us set up a $2M + 1$-nomial model, with M a positive integer. At each time step, one can either have no move (zero jump size) or have possible up or down moves. For a given step size Δ (to be determined later), the allowed down and up jumps will be, respectively,

$$x_j = -j\Delta \text{ and } y_j = j\Delta, \quad j = 1, \ldots, M.$$

Let p be the probability of no jump that is to be determined. Define the conditional probabilities p_j and q_j, $j = 1, \ldots, M$ for a jump of sizes respectively x_j and y_j, $j = 1, \ldots, M$ conditional on a jump by

$$p_j = \frac{A_j}{\sum_{j=1}^{M}(A_j + B_j)},$$

$$q_j = \frac{B_j}{\sum_{j=1}^{M}(A_j + B_j)},$$

where A_j and B_j are derived from the Lévy density as follows:

$$A_j = \int_{-(j+\frac{1}{2})\Delta}^{-(j-\frac{1}{2})\Delta} k(x)\mathrm{d}x$$

$$B_j = \int_{(j-\frac{1}{2})\Delta}^{(j+\frac{1}{2})\Delta} k(x)\mathrm{d}x.$$

This is in line with and inspired by the exact meaning of a Lévy density as explained in Section 2.4. The multinomial characteristic function (over a single time-step) for a given M is given by

$$\phi_M(u) = \phi_M(u; p, \Delta) = p + (1-p)\sum_{j=1}^{M}\left(p_j \exp(\mathrm{i}ux_j) + q_j \exp(\mathrm{i}uy_j)\right).$$

For a given M, having N independent time steps following the above dynamics, the multinomial characteristic function after N time steps, is then simply given by the Nth power of $\phi_M(u; p, \Delta)$. The two parameters p, the probability of no jump, and Δ, the jump spacing, are then to be determined by, for example, a numerical least squares minimization of

$$\text{lse} = \int_{|u|\leq U} \|\phi(u) - (\phi_M(u))^N\|\mathrm{d}u, \tag{2.18}$$

where $U > 0$ is setting a reasonable approximation range.

Example 2.3 *Let us apply the above procedure to approximate a Variance Gamma process. We take* $M = 10$ *and hence have 21 different jump sizes (including a zero jump size). We simulate the process over a one-year horizon using* $N = 50$ *time steps.*

Here we work, as an example, with the stylized annualized parameters $\sigma = 0.2$, $\nu = 0.75$ *and* $\theta = -0.3$. *We perform our least-squared minimization of the error in Equation (2.18) in the range* $|u| \leq U$, *with* $U = 20$.

Assume first $\Delta = 0.01$, $p = 0.3$ *and* $M = 10$. *We recall, see Section 2.5.1, that*

$$k(x) = C\frac{\exp(-Mx)}{x}, \text{ if } x > 0 \tag{2.19}$$

$$= C\frac{\exp(Gx)}{|x|}, \text{ if } x < 0, \tag{2.20}$$

where we work with the CGM*-parametrization of the VG distribution.*

Table 2.3 *Probabilities conditional on a jump.*

j	p_j	q_j
1	0.0154	0.0034
2	0.0177	0.0046
3	0.0206	0.0062
4	0.0244	0.0086
5	0.0296	0.0121
6	0.0368	0.0175
7	0.0478	0.0264
8	0.0664	0.0426
9	0.1045	0.0780
10	0.2334	0.2039

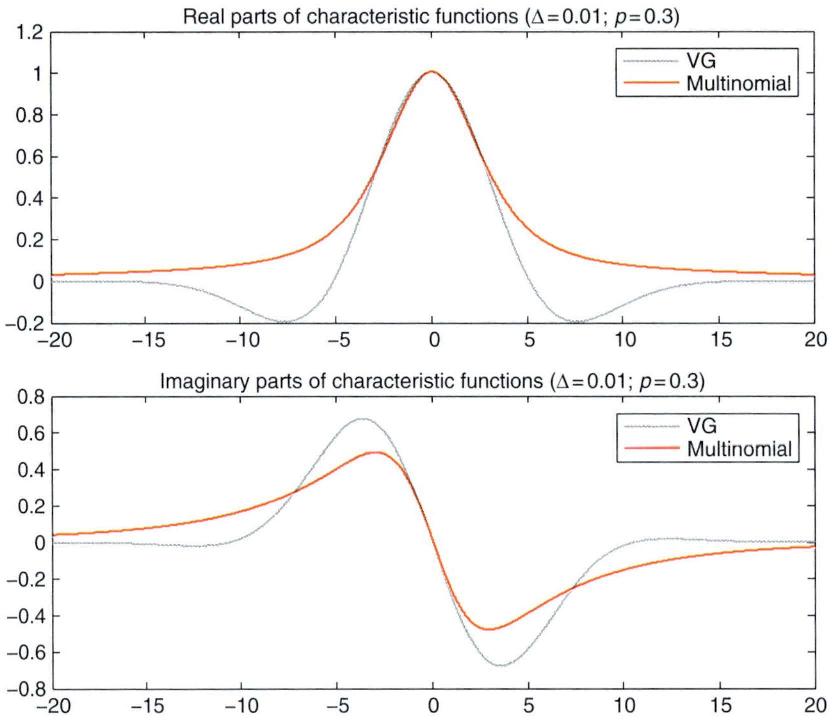

Figure 2.15 Multinomial approximation to VG process (without optimization: $\Delta = 0.01$, $p = 0.30$)

Therefore the A_j and B_j, $j = 1, \ldots, m$, can easily be calculated, and therefore also the p_js and q_js as well (see Table 2.3). However, if we plot, as in Figure 2.15, the real and imaginary parts of the multinomial characteristic function $(\phi_M(u))^N$ and the target VG characteristic function $\phi(u)$, we see that the approximation is not really good yet.

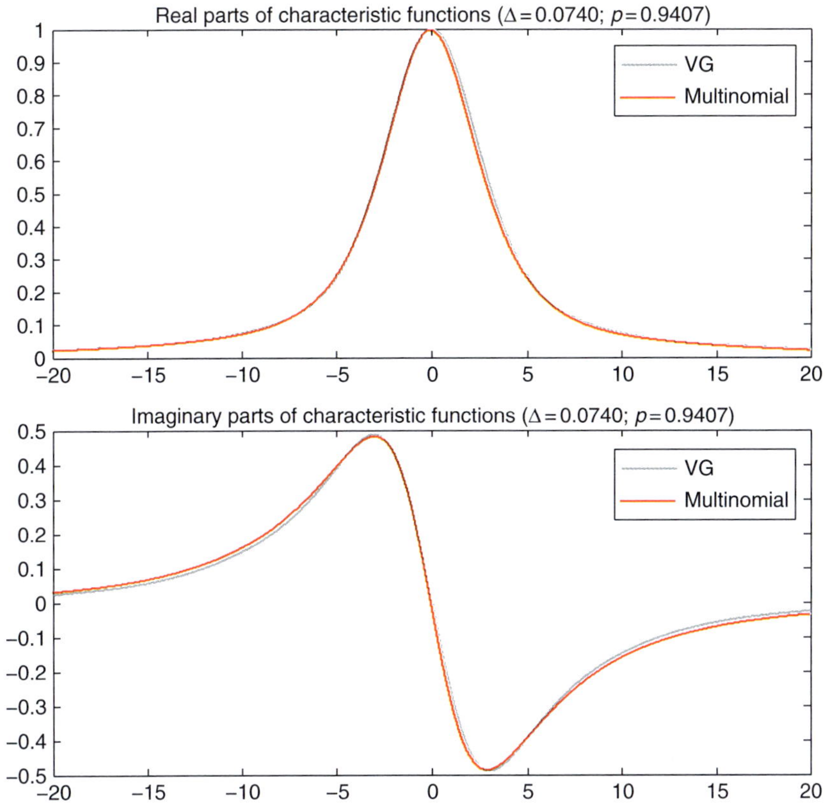

Figure 2.16 Multinomial approximation to VG process (with optimization: $\Delta = 0.0740$, $p = 0.9407$)

In order to improve the approximation, we need to change the Δ *and* p *parameters and minimize the quantity in Equation (2.18). If we perform a numerical optimization minimizing this least-squared error, we obtain a very good fit for* $\Delta = 0.0740$ *and* $p = 0.9407$*, as can be seen in Figure 2.16.*

3
Numerical Techniques

In this chapter we discuss several numerical implementation issues. More precisely, we focus first on several techniques to efficiently calculate the price of a vanilla derivative under advanced models. We use these efficient pricers in a second part in which we focus on the calibration problem: how to determine the (risk-neutral) parameters of a given model, such that we get the best possible match with the dynamics of the underlying asset we are focusing on. Finally, we elaborate on how to perform a Monte Carlo simulation or, in other words, how to simulate several/numerous paths of the underlying under a given model with a given (calibrated) parameter set. This is particularly useful for the pricing of exotic options, where one estimates the price of the exotic as the discounted average payoff over the (many) simulated paths.

3.1 Pricing of Vanilla Options

3.1.1 Pricing Vanilla Options Using Characteristic Functions and FFT Techniques

In this section, we describe how one can very quickly and efficiently price vanilla options using the theory of characteristic functions and Fast Fourier Transforms (FFT). The method is particularly interesting in the case of advanced equity models, like the Variance Gamma model, and many other models (e.g. the Heston model) where no closed-form solutions for vanillas exist.

An important advantage of the method is that the pricer only needs as input the characteristic function of the dynamics of the underlying model. More precisely, the characteristic function of the logarithm of the stock price at the relevant maturity. If one wishes to switch to another model, only the corresponding characteristic functions need to be changed and the actual pricing algorithm remains untouched. The methodology can be applied not only to vanillas, but typically also to more general options which depend only on the stock price at maturity, such as digitals.

Furthermore, a lot of the so-called greeks (Delta, Gamma, ...; see Section 3.1.2) of the classical vanilla options can also be calculated using a similar procedure.

The Carr–Madan Formula

According to the fundamental theorem of asset pricing, the arbitrage free price $EC(K, T)$ of a European Call option with maturity T and strike K is given by

$$EC(K, T) = \exp(-rT)E_Q[(S_T - K)^+],$$

where we take the expectation under Q, i.e. the risk-neutral measure. Carr and Madan (1998) showed the following generic formula:

$$EC(K, T) = \frac{\exp(-\alpha \log(K))}{\pi} \int_0^{+\infty} \exp(-\mathrm{i}v \log(K))\varrho(v)\mathrm{d}v, \tag{3.1}$$

where

$$\varrho(v) = \frac{\exp(-rT)E_Q[\exp(\mathrm{i}(v - (\alpha + 1)\mathrm{i}) \log(S_T))]}{\alpha^2 + \alpha - v^2 + \mathrm{i}(2\alpha + 1)v}, \tag{3.2}$$

where α is a positive constant (see later).

Denote the (risk neutral) characteristic function of the log price process $s_T = \log(S_T)$ at maturity T by

$$\phi(u; T) = E_Q[\exp(\mathrm{i}u \log(S_T))] = E_Q[\exp(\mathrm{i}us_T)].$$

In many situations $\phi(u; T)$ is known analytically: for example, in the Black–Scholes setting; under the Variance Gamma model and more generally under many Lévy models (Schoutens, 2003); under the VG Sato model; under the Heston model (Heston, 1993) or Heston with jumps (Bates) model (Bates, 1996); under certain Lévy models with stochastic volatility (Carr et al., 2003), ... In that case $\varrho(v)$ can be expressed completely analytically. Indeed, the expected value, $E[\exp(\mathrm{i}(v - (\alpha + 1)\mathrm{i}) \log(S_T))]$, in $\varrho(v)$ is nothing other than the characteristic function of the random variable $\log(S_T)$ evaluated in the point $v - (\alpha + 1)\mathrm{i}$:

$$E_Q[\exp(\mathrm{i}(v - (\alpha + 1)\mathrm{i}) \log(S_T))] = \phi(v - (\alpha + 1)\mathrm{i}; T).$$

Example 3.1 *In the Black–Scholes world we have*

$$\begin{aligned}
S_T &= S_0 \exp((r - q - \sigma^2/2)T + \sigma W_T), \textit{ with } W \textit{ standard Brownian motion}\\
s_T &= \log(S_0) + (r - q - \sigma^2/2)T + \sigma W_T\\
s_T &\sim \textit{Normal}(\log(S_0) + (r - q - \sigma^2/2)T, \sigma^2 T)
\end{aligned}$$

$$\phi_{BS}(u; T) = \exp(\mathrm{i}u(\log(S_0) + (r - q - \sigma^2/2)T)) \exp\left(-\frac{1}{2}\sigma^2 T u^2\right).$$

Example 3.2 *In the Variance Gamma world we have*

$$
\begin{aligned}
S_T &= S_0 \exp((r - q + \omega)T + X_T), \textit{ with } X \textit{ is a } VG(C, G, M) \textit{ process}\\
s_T &= \log(S_0) + (r - q + \omega)T + X_T\\
s_T &\sim \log(S_0) + (r - q + \omega)T + VG(CT, G, M)
\end{aligned}
$$

$$
\phi_{VG}(u; T) = \exp(\mathrm{i}u(\log(S_0) + (r - q + \omega)T)) \left(\frac{GM}{GM + (M - G)\mathrm{i}u + u^2} \right)^{CT}.
$$

Using the *Carr–Madan Formula* (3.1) and combining it with numerical techniques that evaluate the integral involved in an efficient way (based on the Fast Fourier Transform (FFT)), one obtains a fast pricing algorithm of the entire option surface. The algorithm generates in one run prices for a fine grid of strikes and all given maturities. Moreover, the formula/algorithm is generic and can be used for any model if the characteristic $\phi(u; T)$ is available. Using this pricing formula/algorithm, a very fast calibration on market option data for advanced models is possible.

Basically, we provide the characteristic function of underlying stochastic dynamics, the model parameters and maturities, as input to our pricing algorithm. The algorithm generates for a whole range of strikes (chosen by the FFT algorithm) and all the given maturities the vanilla prices, as output. If one needs the option price for a particular strike, this is obtained via interpolation. The strike-grid should hence be fine enough to obtain accurate results.

Next, we derive the Carr–Madan formula and show how the Fast Fourier Transform (FFT) can be used to evaluate the integral in the formula.

Let α be a positive constant such that the αth moment of the stock price exists. We comment later on the choice of α. Recall that we suppose that we have explicitly available the characteristic function of $s_T = \log(S_T)$:

$$
\phi(u; T) = E_Q[\exp(\mathrm{i}us_T)] = E_Q[\exp(\mathrm{i}u \log(S_T)].
$$

Denote by $k = \log(K)$ the log-strike, so as we moved from S_T to s_T we also move to log-space for the strikes. If we work in log-space we denote the call price with log-strike k by $C(k, T)$. Assume for simplicity that the (risk-neutral) density function of $s_T = \log(S_T)$ exists and let us denote this density function by $q(x; T)$. Then we have:

$$
\phi(u; T) = \int_{-\infty}^{+\infty} \exp(\mathrm{i}ux) q(x; T) \mathrm{d}x.
$$

We know

$$
\begin{aligned}
C(k, T) &= \exp(-rT) E_Q[(S_T - \mathrm{e}^k)^+]\\
&= \exp(-rT) \int_k^{\infty} (\mathrm{e}^x - \mathrm{e}^k) q(x; T) \mathrm{d}x.
\end{aligned}
$$

However, note that the Call function $C(k, T) \to \exp(-rT)S_0$ (in the log-strike price) as $k \to -\infty$ and is hence not square integrable. It would therefore not be possible to apply Fourier theory. To obtain a square integrable function, we consider the modified call price:

$$c(k, T) = \exp(\alpha k)C(k, T),$$

for some $\alpha > 0$. For a suitable range of positive values for α, we expect $c(k, T)$ to be square integrable in k over the entire real line.

We now show that $\varrho(v)$ of Equation (3.2) is actually the Fourier transform of $c(k, T)$. Indeed

$$\begin{aligned}
&\int_{-\infty}^{+\infty} \exp(\mathrm{i}vk)c(k, T)\mathrm{d}k \\
&= \int_{-\infty}^{+\infty} \exp(\mathrm{i}vk)\exp(\alpha k)C(k; T)\mathrm{d}k \\
&= \int_{-\infty}^{+\infty} \exp(\mathrm{i}vk)\exp(-rT)\exp(\alpha k)\int_{k}^{\infty} (\mathrm{e}^{x} - \mathrm{e}^{k})q(x; T)\mathrm{d}x\mathrm{d}k \\
&= \exp(-rT)\int_{-\infty}^{+\infty} q(x; T)\int_{-\infty}^{x} \exp(\mathrm{i}vk)\exp(\alpha k)(\mathrm{e}^{x} - \mathrm{e}^{k})\mathrm{d}k\mathrm{d}x \\
&= \exp(-rT)\int_{-\infty}^{+\infty} q(x; T)\left(\frac{\exp\left((\alpha + 1 + \mathrm{i}v)x\right)}{\alpha^2 + \alpha - v^2 + \mathrm{i}(2\alpha + 1)v}\right)\mathrm{d}x \\
&= \frac{\exp(-rT)\phi(v - (\alpha + 1)\mathrm{i}; T)}{\alpha^2 + \alpha - v^2 + \mathrm{i}(2\alpha + 1)v} \\
&= \varrho(v).
\end{aligned}$$

Then using the inverse transform we have

$$\begin{aligned}
C(k, T) &= \exp(-\alpha k)c(k; T) \\
&= \exp(-\alpha k)\frac{1}{2\pi}\int_{\infty}^{\infty} \exp(-\mathrm{i}vk)\varrho(v)\mathrm{d}v \\
&= \exp(-\alpha k)\frac{1}{\pi}\int_{0}^{\infty} \exp(-\mathrm{i}vk)\varrho(v)\mathrm{d}v.
\end{aligned}$$

The second equality holds because $C(k, T)$ is real, which implies that the function $\varrho(v)$ is odd in its imaginary part and even in its real part.

Rephrasing gives the Carr–Madan formula:

$$EC(K, T) = C(\log K, T) = \frac{\exp(-\alpha \log(K))}{\pi}\int_{0}^{+\infty} \exp(-\mathrm{i}v\log(K))\varrho(v)\mathrm{d}v,$$

where

$$\varrho(v) = \frac{\exp(-rT)E_Q[\exp(\mathrm{i}(v-(\alpha+1)\mathrm{i})\log(S_T))]}{\alpha^2+\alpha-v^2+\mathrm{i}(2\alpha+1)v}$$
$$= \frac{\exp(-rT)\phi(v-(\alpha+1)\mathrm{i};T)}{\alpha^2+\alpha-v^2+\mathrm{i}(2\alpha+1)v}.$$

Next, we illustrate how the calculation of the call price via the Carr–Madan formula can be done quickly and accurately using the Fast Fourier Transform (FFT). The FFT is an efficient algorithm for computing the following transformation of a vector $(\alpha_n, n = 1, \ldots, N)$ into a vector $(\beta_n, n = 1, \ldots, N)$:

$$\beta_n = \sum_{j=1}^{N} \exp\left(-\frac{\mathrm{i}2\pi(j-1)(n-1)}{N}\right)\alpha_j.$$

Typically N is a power of 2. The number of operations of the FFT algorithm is of the order $\mathcal{O}(N\log N)$, and this in contrast to the straightforward evaluation of the above sums which give rise to $\mathcal{O}(N^2)$ numbers of operations.

An approximation for the integral in the Carr–Madan formula

$$C(k,T) = \exp(-\alpha k)\frac{1}{\pi}\int_0^\infty \exp(-\mathrm{i}vk)\varrho(v)\mathrm{d}v$$

on the N points-grid $(0, \eta, 2\eta, 3\eta, \ldots, (N-1)\eta)$ is

$$C(k,T) \approx \exp(-\alpha k)\frac{1}{\pi}\sum_{j=1}^{N}\exp(-\mathrm{i}v_j k)\varrho(v_j)\eta, \qquad v_j = \eta(j-1).$$

We will calculate the value of these call prices for N log-strikes levels ranging from say $-b$ to b (Note: if $S_0 = 1$, at-the-money corresponds to $b = 0$):

$$k_n = -b + \lambda(n-1), \qquad n = 1, \ldots, N, \qquad \text{where } \lambda = 2b/N.$$

This gives

$$C(k_n,T) \approx \exp(-\alpha k_n)\frac{1}{\pi}\sum_{j=1}^{N}\exp(-\mathrm{i}v_j(-b+\lambda(n-1)))\varrho(v_j)\eta,$$
$$= \exp(-\alpha k_n)\frac{1}{\pi}\sum_{j=1}^{N}\exp(-\mathrm{i}\eta\lambda(j-1)(n-1))\exp(\mathrm{i}v_j b)\varrho(v_j)\eta.$$

If we choose λ and η such that $\lambda\eta = 2\pi/N$, then

$$C(k_n,T) \approx \exp(-\alpha k_n)\frac{1}{\pi}\sum_{j=1}^{N}\exp\left(-\frac{\mathrm{i}2\pi(j-1)(n-1)}{N}\right)\exp(\mathrm{i}v_j b)\varrho(v_j)\eta.$$

The summation in the approximation above is an exact application of the FFT on the vector $(\exp(iv_j b)\varrho(v_j)\eta, \ j = 1, \ldots, N)$. Note that, by fixing $\lambda\eta = 2\pi/N$, taking a smaller grid-size η makes the grid-size λ (for the log-strike grid) larger. Carr and Madan (1999) report that the following choice gave very satisfactory results:

$$\eta = 0.25$$
$$N = 4096$$
$$\alpha = 1.5,$$

which implies

$\lambda = 0.0061$ or an inter-strike range a little over a half a percentage

$b = 12.57.$

A more refined weighting (Simpson's rule) for the integral in the Carr–Madan formula on the N points-grid $(0, \eta, 2\eta, 3\eta, \ldots, (N - 1)\eta)$ leads to the more accurate following approximation

$$C(k, T) \approx \exp(-\alpha k)\frac{1}{\pi}\sum_{j=1}^{N}\exp(-iv_j k)\varrho(v_j)\eta\left(\frac{3 + (-1)^j - \delta_{j-1}}{3}\right),$$
$$v_j = \eta(j - 1),$$

where $\delta_j = 1$, if $j = 0$ and $\delta_j = 0$, if $j \neq 0$.

3.1.2 Calculation of the Greeks Using FFT Techniques

To analyse risks involved in a particular option, one often calculates the partial derivatives of the price of the option with respect to its parameters. These partial derivatives are commonly known as *greeks*. Below, we detail the calculations for the Delta and the Gamma.

Delta

The *Delta* of an option, often denoted by Δ, measures the sensitivity of the option's value to price changes in the underlying.

$$\Delta = \frac{\partial EC(K, T)}{\partial S_0}.$$

In the Black–Scholes setting this Delta is giving the number of stocks one needs to hold in order to perfectly replicate the option. In the more advanced models, perfect replication is no longer always possible and many other hedging strategies can be considered and maybe make more sense. Nevertheless, Delta-hedging is a very well accepted strategy that is also straightforward to apply. We must note, however, that taking (partial) derivatives is a local operator, typically taking into account continuous movements of the underlying. If we work with models in which the underlying price process can jump, such a local operator does not tell the whole story.

We have:

$$\begin{aligned}\Delta &= \frac{\partial}{\partial S_0}\left[\frac{\exp(-\alpha\log(K))}{\pi}\int_0^{+\infty}\exp(-\mathrm{i}v\log(K))\frac{\exp(-rT)\phi(v-(\alpha+1)\mathrm{i};T)}{\alpha^2+\alpha-v^2+\mathrm{i}(2\alpha+1)v}\mathrm{d}v\right]\\ &= \frac{\exp(-\alpha\log(K))}{\pi}\int_0^{+\infty}\exp(-\mathrm{i}v\log(K))\frac{\exp(-rT)\frac{\partial\phi(v-(\alpha+1)\mathrm{i};T)}{\partial S_0}}{\alpha^2+\alpha-v^2+\mathrm{i}(2\alpha+1)v}\mathrm{d}v.\end{aligned}$$

Now, assume the (risk-neutral) model for the price of the underlying at time T is of the form $S_T = S_0\bar{S}_T$, where $\bar{S}_T$ is not depending on S_0 anymore, then $\log(S_T) = \log(S_0) + \log(\bar{S}_T)$. Note that this assumption is a very typical one and is the case for all models we consider. Furthermore, we then have that

$$\begin{aligned}\frac{\partial\phi(v-(\alpha+1)\mathrm{i};T)}{\partial S_0} &= \frac{\partial}{\partial S_0}E_Q[\exp(\mathrm{i}(v-(\alpha+1)\mathrm{i})(\log(S_0)+\log(\bar{S}_T))]\\ &= \frac{\phi(v-(\alpha+1)\mathrm{i};T)(\alpha+1+\mathrm{i}v)}{S_0}.\end{aligned}$$

In conclusion,

$$\begin{aligned}\Delta &= \frac{\exp(-\alpha\log(K))}{\pi}\\ &\quad\int_0^{+\infty}\exp(-\mathrm{i}v\log(K))\frac{\exp(-rT)\phi(v-(\alpha+1)\mathrm{i};T)(\alpha+1+\mathrm{i}v)}{S_0(\alpha^2+\alpha-v^2+\mathrm{i}(2\alpha+1)v)}\mathrm{d}v;\\ &= \frac{\exp(-\alpha\log(K))}{\pi}\int_0^{+\infty}\exp(-\mathrm{i}v\log(K))\frac{\exp(-rT)\phi(v-(\alpha+1)\mathrm{i};T)}{S_0(\alpha+\mathrm{i}v)}\mathrm{d}v,\end{aligned}$$

where we used in the last line the fact that $\alpha^2+\alpha-v^2+\mathrm{i}(2\alpha+1)v = (\alpha+\mathrm{i}v)(\alpha+1+\mathrm{i}v)$.

Gamma

The *Gamma* of an option, often denoted by Γ, measures the sensitivity of the Delta of the option with respect to price changes in the underlying.

$$\Gamma = \frac{\partial\Delta}{\partial S_0} = \frac{\partial^2 EC(K,T)}{\partial S_0^2}.$$

High Gammas mean that the corresponding Delta-hedge position is very sensitive to changes in the underlying price process.

Completely analogous as in the calculation of Delta, we have

$$\Gamma = \frac{\exp(-\alpha \log(K))}{\pi} \int_0^{+\infty} \exp(-\mathrm{i}v \log(K)) \frac{\exp(-rT) \frac{\partial^2 \phi(v-(\alpha+1)\mathrm{i};T)}{\partial S_0^2}}{\alpha^2 + \alpha - v^2 + \mathrm{i}(2\alpha + 1)v} \mathrm{d}v.$$

Under the same assumption on the form of the price process, we then have that

$$\frac{\partial}{\partial S_0} \frac{\phi(v - (\alpha+1)\mathrm{i}; T)}{S_0} = \frac{\phi(v - (\alpha+1)\mathrm{i}; T)(\alpha + \mathrm{i}v)}{S_0^2}.$$

Hence, in conclusion

$$\Gamma = \frac{\exp(-\alpha \log(K))}{\pi} \int_0^{+\infty} \exp(-\mathrm{i}v \log(K)) \frac{\exp(-rT)\phi(v - (\alpha+1)\mathrm{i}; T)}{S_0^2} \mathrm{d}v.$$

3.1.3 Quadrature Techniques and the VG Laguerre Model

In this section, following the idea proposed in Madan et al. (2013) (see also Guillaume (2011)), we compute VG option prices as a weighted sum of Black–Scholes option prices by making use of the Gauss–Laguerre quadrature. The implementation results in an extremely fast pricing algorithm. However, the method is not generic, as is the above Carr–Madan Formula, and very much relies on exact properties of the VG distribution.

Gauss–Laguerre Quadrature

In numerical analysis, the Gauss–Laguerre quadrature is an extension of the Gaussian quadrature method for approximating the value of integrals of the following kind:

$$\int_0^{+\infty} \exp(-z) f(z) \mathrm{d}z.$$

In this case the integral is approximated by

$$\int_0^{+\infty} \exp(-z) f(z) \mathrm{d}z \approx \sum_{i=1}^{N} w_i f(z_i),$$

and z_i is the ith root of the Laguerre polynomial $L_N(z)$ and the weight w_i can be expressed as

$$w_i = \frac{z_i}{((N+1)L_{N+1}(z_i))^2}, \quad i = 1, \ldots, N. \tag{3.3}$$

The Laguerre polynomial of degree n is given by

$$L_n(z) = \sum_{k=0}^{n} \binom{n}{k} \frac{(-1)^k}{k!} z^k.$$

Table 3.1 *Gauss–Laguerre weights and node points for $N = 5$.*

i	z_i	w_i
1	0.263560319718141	0.521755610582808
2	1.413403059106517	0.398666811083176
3	3.596425771040722	0.075942449681707
4	7.085810005858837	0.003611758679922
5	12.640800844275782	0.000023369972386

Table 3.2 *Gauss–Laguerre weights and node points for $N = 10$.*

i	z_i	w_i
1	0.137793470540492	0.308441115765020
2	0.729454549503170	0.401119929155274
3	1.808342901740317	0.218068287611809
4	3.401433697854900	0.062087456098678
5	5.552496140063804	0.009501516975181
6	8.330152746764497	0.000753008388587
7	11.843785837900066	0.000028259233496
8	16.279257831378104	0.000000424931398
9	21.996585811980761	0.000000001839565
10	29.920697012273891	0.000000000000991

For a fix N, we thus need to calculate the zeros z_i, $i = 1, \ldots, N$, of $L_N(z)$. These zeros need to be calculated numerically and from these the weights w_i, $i = 1, \ldots, N$, can be calculated using Equation (3.3).

Example 3.3 *For $N = 5$ and $N = 10$, the nodes z_i and weights x_i, $i = 1, \ldots, N$, are given in Tables 3.1 and 3.2.*

The VG Laguerre Model

We now show how the cumulative distribution function (cdf) $F_{VG}(x; \sigma, \nu, \theta)$ of a $\mathbf{VG}(\sigma, \nu, \theta)$-random variable X can be obtained by performing a Gauss–Laguerre quadrature. We have that

$$X =^{law} \theta G + \sigma \sqrt{G} Z,$$

with G a Gamma$(1/\nu, 1/\nu)$ random variable and Z an independent standard normal random variable. Therefore, we can write

$$\begin{aligned} F_{VG}(x; \sigma, \nu, \theta) &= P(\theta G + \sigma \sqrt{G} Z \leq x) \\ &= \int_0^{+\infty} P(\theta g + \sigma \sqrt{g} Z \leq x) f_{Gamma(1/\nu, 1/\nu)}(g) \mathrm{d}g \end{aligned}$$

$$
\begin{aligned}
&= \int_0^{+\infty} P(\theta g + \sigma\sqrt{g}Z \le x)\frac{(1/\nu)^{1/\nu}}{\Gamma(1/\nu)} g^{\frac{1}{\nu}-1}\exp(-g/\nu)\mathrm{d}g \\
&= \int_0^{+\infty} P\left(Z \le \frac{x-\theta z\nu}{\sigma\sqrt{z\nu}}\right)\frac{z^{\frac{1}{\nu}-1}}{\Gamma(1/\nu)}\exp(-z)\mathrm{d}z \\
&\approx \sum_{i=1}^{N}\frac{z_i^{\frac{1}{\nu}-1}}{\Gamma(1/\nu)}F_{N(0,1)}\left(\frac{x-\theta\nu z_i}{\sigma\sqrt{\nu z_i}}\right)w_i,
\end{aligned}
$$

where $F_{N(0,1)}$ is the cdf of the standard Normal distribution, and the z_is and the w_is denote the roots and weights as defined above, respectively. We will replace $\Gamma\left(\frac{1}{\nu}\right)$ in order to insure that we have a proper cdf with

$$
\sum_{i=1}^{N} z_i^{\frac{1}{\nu}-1} w_i.
$$

We hence conclude that

$$
F_{VG}(x;\sigma,\nu,\theta) \approx \left(\frac{1}{\sum_{i=1}^{N} z_i^{\frac{1}{\nu}-1} w_i}\right)\sum_{i=1}^{N} z_i^{\frac{1}{\nu}-1} F_{N(0,1)}\left(\frac{x-\theta\nu z_i}{\sigma\sqrt{\nu z_i}}\right) w_i. \tag{3.4}
$$

Next, we will use the above approximation of the VG cdf for option pricing. Although the approximation is quite good, our stock price model under this approximation will no longer be perfectly risk-neutral. The variable taking care of this risk-neutrality is ω and it is calculated (see Section 2.5.3) as

$$
\omega = \frac{-\log\left(E_Q[\exp(X_T)]\right)}{T}.
$$

Under our approximation we now have that

$$
\begin{aligned}
&E_Q[\exp(X_T)] \\
&= \int_{-\infty}^{+\infty} \exp(x)\,\mathrm{d}F_{VG}(x;\sigma\sqrt{T},\nu/T,\theta T) \\
&\approx \left(\frac{1}{\sum_{i=1}^{N} z_i^{\frac{T}{\nu}-1} w_i}\right)\sum_{i=1}^{N} z_i^{\frac{T}{\nu}-1} w_i\left(\int_{-\infty}^{+\infty}\exp(x)\,\mathrm{d}F_{N(0,1)}\left(\frac{x-\theta\nu z_i}{\sigma\sqrt{\nu z_i}}\right)\right) \\
&= \left(\frac{1}{\sum_{i=1}^{N} z_i^{\frac{T}{\nu}-1} w_i}\right)\sum_{i=1}^{N} z_i^{\frac{T}{\nu}-1} w_i \exp\left(\nu z_i\sigma^2/2 + \theta\nu z_i\right).
\end{aligned}
$$

Note that now ω has become maturity dependent and we see some slight variation over maturity making the appropriate corrections to ensure perfect risk-neutrality at each maturity.

The price $EC(K,T)$ of a European Call option with maturity T and strike K is given by

$$
\begin{aligned}
&EC(K,T)\\
&= \exp(-rT)E_Q[(S_T-K)^+]\\
&= \exp(-rT)E_Q[(S_0\exp((\omega+r-q)T+X_T))-K)^+]\\
&= \exp(-rT)\int_{-\infty}^{+\infty}(S_0\exp\left((\omega+r-q)T+x\right)-K)^+\,\mathrm{d}F_{VG}(x;\sigma\sqrt{T},\nu/T,\theta T)\\
&\approx \frac{\exp(-rT)}{\sum_{i=1}^{N} z_i^{\frac{T}{\nu}-1}w_i}\sum_{i=1}^{N} z_i^{\frac{T}{\nu}-1}w_i\\
&\left(\int_{-\infty}^{+\infty}(S_0\exp\left((\omega+r-q)T+x\right)-K)^+\,\mathrm{d}F_{N(0,1)}\left(\frac{x-\theta\nu z_i}{\sigma\sqrt{\nu z_i}}\right)\right)\\
&= \frac{\exp(-rT)}{\sum_{i=1}^{N} z_i^{\frac{T}{\nu}-1}w_i}\sum_{i=1}^{N} z_i^{\frac{T}{\nu}-1}w_i\\
&\left(\int_{-\infty}^{+\infty}\left(S_0\exp\left((\omega+r-q)T+\theta\nu z_i+\sigma\sqrt{\nu z_i}x\right)-K\right)^+\,\mathrm{d}F_{N(0,1)}(x)\right).
\end{aligned}
$$

The above equation can be seen as a weighted sum of Black–Scholes call options:

$$EC(K,T)=\sum_{i=1}^{N} w_i^{\mathrm{BS}} EC_i^{\mathrm{BS}}(K,T),$$

where $EC_i^{\mathrm{BS}}(K,T)$, $i=1,\dots,N$, is the Black–Scholes price of a European Call option with strike price K and maturity T and corresponding volatility σ_i^{BS}, dividend yield q_i^{BS} (r remains unchanged):

$$
\begin{aligned}
\sigma_i^{\mathrm{BS}} &= \frac{\sigma\sqrt{\nu z_i}}{\sqrt{T}}\\
q_i^{\mathrm{BS}} &= q-\frac{\theta\nu z_i}{T}-\omega-\frac{1}{2}\left(\sigma_i^{\mathrm{BS}}\right)^2\\
w_i^{\mathrm{BS}} &= \frac{z_i^{\frac{T}{\nu}-1}w_i}{\sum_{i=1}^{N} z_i^{\frac{T}{\nu}-1}w_i}.
\end{aligned}
$$

Note that for put prices we have exactly the same weighting scheme but now with put prices calculated under the Black–Scholes setting:

$$EP(K,T)=\sum_{i=1}^{N} w_i^{\mathrm{BS}} EP_i^{\mathrm{BS}}(K,T),$$

Table 3.3 *VG price as weighted Black–Scholes prices for* $N = 10$.

i	q_i^{BS}	σ_i^{BS}	$EC_i^{\mathrm{BS}}(K,T)$	w_i^{BS}
1	−0.232199847060553	0.043921618681088	63.026934105532064	0.004416219920395
2	−0.174217061322210	0.101056240247915	45.622407595839846	0.126851675267692
3	−0.068486002802970	0.159112540751395	21.497464933951775	0.372266108279531
4	0.087636895216259	0.218220236847934	5.992166412337816	0.342633626045531
5	0.298441014552732	0.278809874216989	1.009246378516560	0.130277022762445
6	0.570651362009400	0.341499836683274	0.101317470016857	0.021929953175955
7	0.914987404940685	0.407201426483996	0.005654965500743	0.001582120250747
8	1.349663660301533	0.477398795179977	0.000152398641502	0.000042948814504
9	1.909961802400594	0.554934411771094	0.000001492865848	0.000000325172959
10	2.686524700029320	0.647216932853147	0.000000002547571	0.000000000310240

where w_i^{BS} is the same weight as above and where $EP_i^{\mathrm{BS}}(K,T)$, $i = 1, \ldots, N$, is the Black–Scholes price of a European Put option with strike price K and maturity T using an interest rate r, a dividend yield q_i^{BS} and a volatility σ_i^{BS} also as above.

The approximation formulae lead to a fast computation of European option prices under both the Lévy and Sato settings, since it just requires the valuation of N Black–Scholes option prices which are directly obtained by their celebrated closed-form formula.

Example 3.4 *We calculate the price of an ATM European Call option under the VG model for* $T = 2$, $K = S_0 = 100$, $r = 0.02$, $q = 0.01$, $\sigma = 0.2$, $\theta = -0.3$ *and* $\nu = 0.7$ *for a quadrature with* $N = 10$ *points. The weights and nodes for this quadrature can be found in Table 3.2. In Table 3.3, we provide* q_i^{BS} *and* σ_i^{BS} *and corresponding* $EC_i^{BS}(K,T)$*. These final prices are then weighted using the* w_i^{BS}*. Taking the weighted sum (compare with the final two columns of Table 3.3) yields 16.255, which compares with a price calculated under the FFT framework of 16.258.*

3.2 Calibration

A calibration procedure looks for the optimal parameter set such that model prices match as well as possible the market prices. We elaborate on two methods. A method which tries to minimize the root-mean-square error (or any other error measure) between the observed option prices in the market and corresponding option prices calculated under the model by varying the parameters over the parameter space. Another method, which is called the implied moment matching calibration, infers from the observed market option prices the moments (to some degree) of the return distribution and then looks for the parameter set giving rise to the same moments. The first method typically employs a search or optimization algorithm, needs a starting value and has the potential problem of returning not the global optimal set, but rather a set reflecting a local minimum. The advantage of the moment matching methodology is that no search algorithm or starting value is needed and the problem of a local minimum is avoided. For the VG case, we even have a closed-form solution available. One can infer from option data at a given maturity the optimal parameters almost instantaneously. The disadvantage is that the method really works well only on a single maturity, and moreover that for certain combinations of the moments inferred from the market, no parameter set exists that gives rise to a distribution with exactly these moments; hence, certain approximations need to be made.

3.2.1 Calibration on Vanilla Option Prices via Optimization Algorithms

By calling the pricing algorithm many times (for the maturities available in the data), an optimization procedure can search the parameter space (starting at the initial guess) looking for an "optimal" parameter set. The optimization algorithm tries to minimize the error between model prices and market prices. One has different choices of error functions that one can minimize. The most common one is the root-mean-square error (rmse):

$$\text{rmse} = \sqrt{\frac{1}{N}\sum_{i=1}^{N}\left(\text{Model price}_i - \text{Market price}_i\right)^2},$$

where N is the number of options in the calibration set.

A good measure to compare fits is the average pricing error (ape), defined as

$$\text{ape} = \frac{\sum_{i=1}^{N} |\text{Model price}_i - \text{Market price}_i|}{\sum_{i=1}^{N} \text{Market price}_i}.$$

As output, one obtains essentially "optimal" parameters for which the related model prices fit the market prices best (in, for example, the rmse sense).

There are also many possible search/optimization algorithms that one can deploy. An example is the Nelder–Mead simplex algorithm, which one can use for non-linear unconstrained optimization problems. Further, there is also a choice of which options to include in a set of calibration instruments. Put–call parity shows that calls and puts with the same maturity and same strike are linked to each other, and having one implies having the other. We choose to perform the calibration on OTM puts and OTM calls only. Including only OTM options has its advantages: OTM options are usually more liquid and hence prices are a bit more reliable. Furthermore, OTM options only have time value and no intrinsic value. Since the intrinsic value is deterministic, it contains no usual information and the focus of option pricing is on the time value. In addition, the options with the highest absolute prices are around the current spot. Since the root-mean-squared error function basically compares absolute prices, the highest relative accuracy will be achieved in that region.

It is well known that the Black–Scholes model cannot fit (using a single volatility parameter) a whole option surface. That is the reason why one typically has non-flat implied volatility surfaces (see Section 2.2.6) using basically, for each data point, a different volatility parameter to obtain a fit. The VG model has three parameters and has hence a bit more flexibility to fit an option surface. However, in case one wants to calibrate jointly on a multiple of maturities, this typically also does not lead to an accurate fit. Using Lévy models, one is actually only able to obtain in most cases a good fit on a single maturity. This drawback was actually one of the reasons for looking to more sophisticated models like the Sato models. The VG Sato model has four parameters and is capable of fitting surfaces with a multiple of maturities to a reasonable degree in most cases.

Example 3.5 *As an illustration, we show the fits for the Sato VG model in Figures 3.1 and 3.2 on S&P 500 options for two dates, namely 12 December 2014 and 17 July 2015. We minimized the rmse and used* $\sigma = 0.2, \nu = 0.7, \theta = -0.3$ *and* $\gamma = 0.5$ *as our initial guess.*

The first data set contained 606 option prices. The optimal parameter set results in an rmse of 1.1695; the ape equals 3.49%. The optimal parameters are $\sigma = 0.1720, \nu = 1.4626, \theta = -0.1398, \gamma = 0.5680.$ *The second data set contained 487 prices. The rmse equals 0.7734 and the ape equalled 3.63%; the optimal parameters are* $\sigma = 0.1331, \nu = 1.5360, \theta = -0.1129, \gamma = 0.7061.$

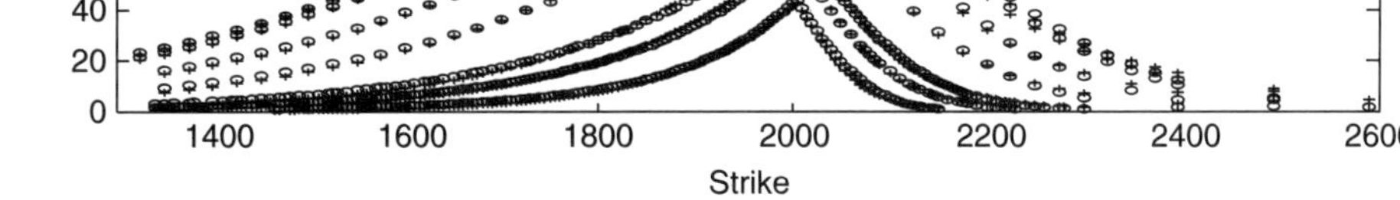

Figure 3.1 SP500 – 12 December 2014 – VG Sato ($\sigma = 0.1720$, $\nu = 1.4626$, $\theta = -0.1398$, $\gamma = 0.5680$)

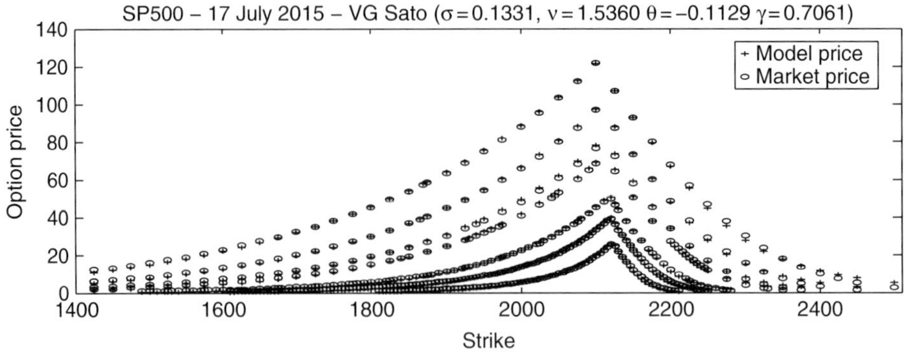

Figure 3.2 SP500 – 17 July 2015 – VG Sato ($\sigma = 0.1331, \nu = 1.5360, \theta = -0.1129, \gamma = 0.7061$)

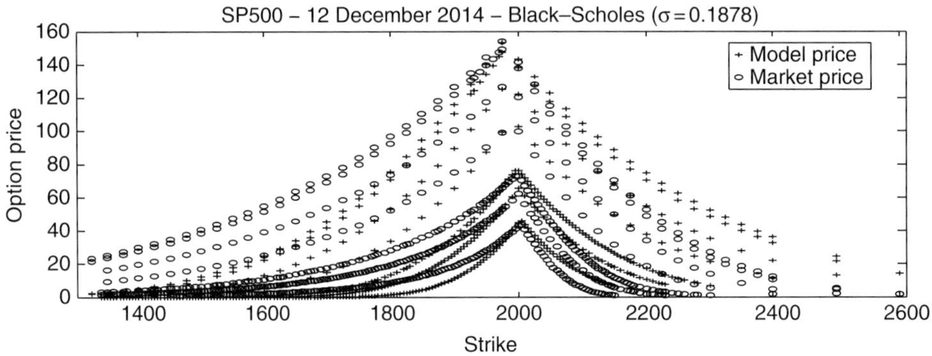

Figure 3.3 SP500 – 12 December 2014 – Black–Scholes ($\sigma = 0.1878$)

If we were to calibrate the Black–Scholes model on these dates, the fit would be much worse. On the first date, a $\sigma = 0.1878$ *leads to the "best" fit with an rmse equal to 12.7085 and an ape of 35.93%. In Figure 3.3 one sees that the Black–Scholes model is producing prices quite off the market prices.*

A calibration of the VG model results in a better, but not really satisfying, fit. On the first date, the parameter combination $\sigma = 0.1512$*,* $\nu = 0.6465$ *and* $\theta = -0.2045$ *leads to a rmse equal to 4.2138 and an ape of 10.09%. In Figure 3.4 one sees that the VG model is producing prices more in line with market prices than the Black–Scholes model, but is still significantly off compared with the VG Sato fit.*

Further, we also show in Figures 3.5–3.8 the evolution of the optimal VG Sato parameters resulting from such a calibration exercise on a daily basis over a period of about 10 years. One clearly sees that parameters change over time, which is natural since the market perception of risk changes over time: there are low volatile

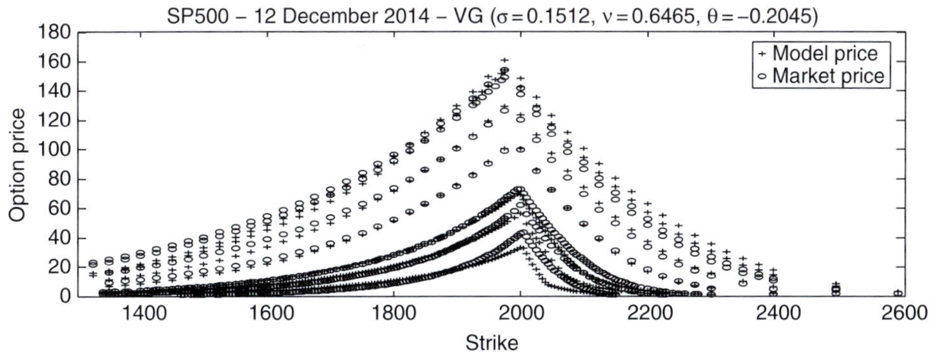

Figure 3.4 SP500 – 12 December 2014 – VG ($\sigma = 0.1512$, $\nu = 0.6465$ and $\theta = -0.2045$)

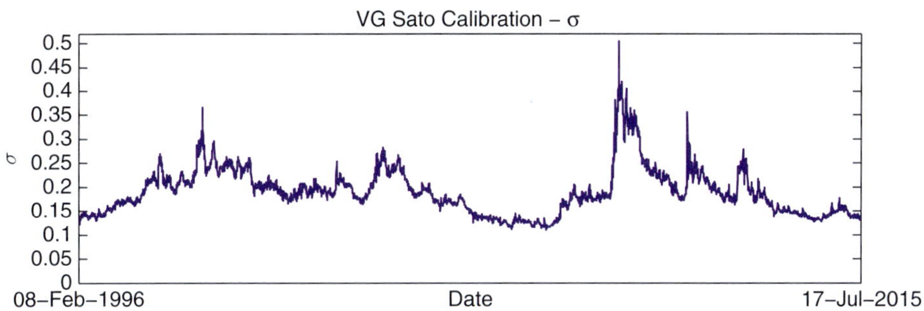

Figure 3.5 Sato VG calibration: σ-time series

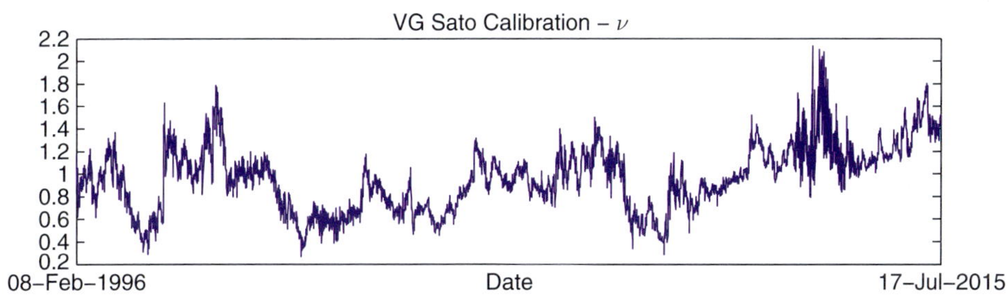

Figure 3.6 Sato VG calibration: ν-time series

periods and high volatile periods, distressed markets and calm markets, skewed views and less skewed views, …

In the sequel, we will repeatedly use the VG model as illustration in the examples. We often choose the parameters equal to $\sigma = 0.2$, $\nu = 0.75$ and $\theta = -0.3$. Noting that the VG Sato model with parameters $(\sigma, \nu, \theta, \gamma)$ has at a unit time a VG distribution with parameters (σ, ν, θ), from the above pictures it can be seen that such a choice is not an unreasonable parameter set to serve for the illustrations;

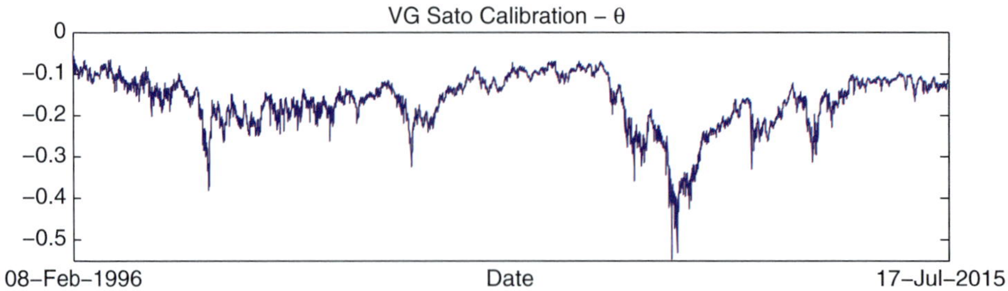

Figure 3.7 Sato VG calibration: θ-time series

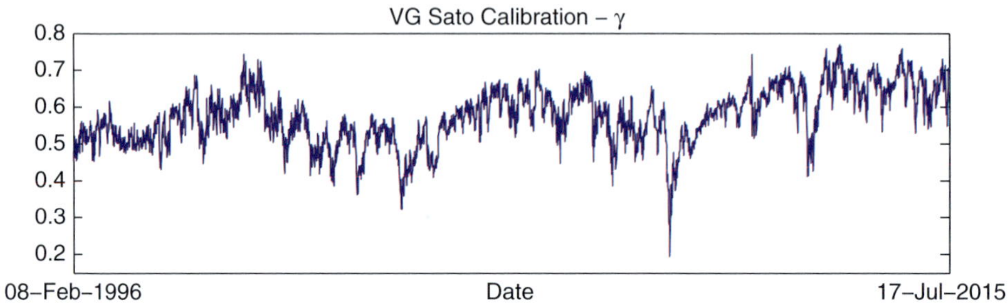

Figure 3.8 Sato VG calibration: γ-time series

of course, in real live examples, one would recalibrate the parameters periodically (weekly, daily, intra-day, ...) and one would employ the latest "optimal" set each time.

3.2.2 Implied Moment Matching Calibration

For certain models, a completely alternative calibration approach can be followed. It avoids optimization algorithms and the related problems that accompany them: optimization algorithms typically need an initial parameter guess, and the choice of this can influence the computation time greatly; and, in addition, different choices can sometimes lead to different output. This is due to the fact that, for many optimization algorithms, there is no guarantee that the output parameters are really "optimal". It could well be that the parameters basically result from a local minimum of the error function, which is not necessarily the global minimum of this error function. The following method tries to match as well as possible, the moments of the distribution with corresponding moments estimated from option prices.

Bakshi and Madan (2000) and Carr and Madan (2001) have shown that any twice-differentiable payoff can be expressed in the form:

$$f(S_T) = f(\kappa) + f'(\kappa)\left((S_T-\kappa)^+ - (\kappa - S_T)^+\right) \tag{3.5}$$
$$+ \int_0^{\kappa} f''(K)(K-S_T)^+ \mathrm{d}K + \int_{\kappa}^{\infty} f''(K)(S_T-K)^+ \mathrm{d}K,$$

with κ a certain strike value.

Taking the expectation of Equation (3.5), we have

$$E_Q\left[f(S_T)\right] = f(\kappa) + f'(\kappa)\exp(rT)\left(EC(\kappa,T) - EP(\kappa,T)\right) \tag{3.6}$$
$$+\exp(rT)\left(\int_0^{\kappa} f''(K)EP(K,T)\mathrm{d}K + \int_{\kappa}^{\infty} f''(K)EC(K,T)\mathrm{d}K\right),$$

where $EC(K,T)$ and $EP(K,T)$ denote the price of a European Call and a Put option with time to maturity T and strike price K, respectively. Making used of the put–call parity

$$EC(K,T) - EP(K,T) = \exp(-qT)S_0 - \exp(-rT)K, \tag{3.7}$$

one can write

$$E_Q\left[f(S_T)\right] = f(\kappa) + f'(\kappa)\left(\exp((r-q)T)S_0 - \kappa\right)$$
$$+\exp(rT)\left(\int_0^{\kappa} f''(K)EP(K,T)\mathrm{d}K + \int_{\kappa}^{\infty} f''(K)EC(K,T)\mathrm{d}K\right).$$

Hence we can derive a closed-form expression for the Nth moment of the log asset-return $X_T = \log\left(\frac{S_T}{S_0}\right)$ by considering the function $f(S_T) = \left(\log\left(\frac{S_T}{S_0}\right)\right)^N$. More precisely, with $F_0 = S_0\exp((r-q)T)$ we have

$$E_Q\left[\left(\log\left(\frac{S_T}{S_0}\right)\right)^N\right] \tag{3.8}$$
$$= \left(\log\left(\frac{\kappa}{S_0}\right)\right)^N + N\left(\log\left(\frac{\kappa}{S_0}\right)\right)^{N-1}\left(\frac{F_0}{\kappa} - 1\right)$$
$$+\exp(rT)\int_0^{\kappa}\frac{N}{K^2}\left((N-1)\left(\log\left(\frac{K}{S_0}\right)\right)^{N-2} - \left(\log\left(\frac{K}{S_0}\right)\right)^{N-1}\right)EP(K,T)\mathrm{d}K$$
$$+\exp(rT)\int_{\kappa}^{\infty}\frac{N}{K^2}\left((N-1)\left(\log\left(\frac{K}{S_0}\right)\right)^{N-2} - \left(\log\left(\frac{K}{S_0}\right)\right)^{N-1}\right)EC(K,T)\mathrm{d}K.$$

Since, in practice, European vanilla options are traded for a discrete set of strikes and not for a continuum of strikes, the continuum of options will be approximated by a finite-weighted sum of vanilla options. From Equation (3.9), we can thus extract a model-independent approximation of the Nth moment from quoted European vanilla option prices. Indeed, we can approximate the integral of the right-hand side with respect to a trapezoidal scheme, where the integral is truncated beforehand on the range of listed strikes:

$$E_Q\left[\left(\log\left(\frac{S_T}{S_0}\right)\right)^N\right] = \left(\log\left(\frac{K_0}{S_0}\right)\right)^N + N\left(\log\left(\frac{K_0}{S_0}\right)\right)^{N-1}\left(\frac{F_0}{K_0}-1\right)$$
$$+\exp(rT)N\sum_{i=1}^{M}\frac{\Delta K_i}{K_i^2}\left((N-1)\left(\log\left(\frac{K_i}{S_0}\right)\right)^{N-2}\right.$$
$$\left.-\left(\log\left(\frac{K_i}{S_0}\right)\right)^{N-1}\right)\text{Price}(K_i, T), \quad (3.9)$$

and where:

- F_0 is the current forward price: $F_0 = F_0(T) = S_0 \exp((r-q)T)$;
- T is the option maturity in years;
- r is the risk-free interest rate corresponding to the maturity T;
- K_i is the strike price of the ith out-of-the-money option (at maturity T); (An out-of-the-money call option is characterized by $K_i > F_0$ and an out-of-the-money put option by $K_i < F_0$.)
- ΔK_i is the interval between the strikes:

$$\begin{cases} \Delta K_1 &= K_2 - K_1, \\ \Delta K_i &= \frac{K_{i+1}-K_{i-1}}{2}, \quad \forall i \neq 1 \text{ or } N \\ \Delta K_N &= K_N - K_{N-1}; \end{cases}$$

- K_0 is the first listed strike below the forward stock index level F_0;
- $\text{Price}(K_i, T)$ for the option with strike K_i and maturity T is given by

$$\begin{cases} \text{Price}(K_i, T) = EP(K_i, T) & \text{if } K_i < K_0 \\ \text{Price}(K_i, T) = \frac{EC(K_0,T)+EP(K_0,T)}{2} & \text{if } K_i = K_0 \\ \text{Price}(K_i, T) = EC(K_i, T) & \text{if } K_i > K_0 \end{cases}$$

  where the option prices $EC(K, T)$ and $EP(K, T)$ are taken to be equal to the mid-point of the bid and the ask.

We can then infer the market-implied variance, skewness and kurtosis of the log-asset return $X_T = \log(S_T/S_0)$ by

$$\begin{cases} v = \text{Var}^{\text{Market}}(X_T) & = E_Q\left[X_T^2\right] - \left(E_Q\left[X_T\right]\right)^2 \\ s = \text{Skewness}^{\text{Market}}(X_T) & = \frac{E_Q[X_T^3]-3E_Q[X_T]E_Q[X_T^2]+2\left(E_Q[X_T]\right)^3}{(\text{Var}(X_T))^{3/2}} \\ k = \text{Kurtosis}^{\text{Market}}(X_T) & = \frac{E_Q[X_T^4]-4E_Q[X_T]E_Q[X_T^3]+6\left(E_Q[X_T]\right)^2E_Q[X_T^2]-3\left(E_Q[X_T]\right)^4}{(\text{Var}(X_T))^2}, \end{cases} \quad (3.10)$$

where the moments of the log asset return $E_Q\left[X_T^N\right]$ are computed by making use of the approximations (3.9). Moreover, if the characteristic function of the log-asset return X_T is known in closed form under the particular model we consider, we can directly derive a model-dependent closed-form expression for the Nth moment

$$E_Q\left[X_T^N\right] = \mathrm{i}^{-N}\frac{d^N}{du^N}\phi_{X_T}(u)|_{u=0}, \tag{3.11}$$

assuming that $E_Q\left[|X_T|^N\right] < \infty$. Since $\phi_{X_T}(u)$ is available for a wide range of models proposed in the financial literature, we can then infer the N model parameters by matching the second to the $(N+1)$th market and model implied moments, or equivalently the second to the $(N+1)$th standardized moments $E_Q\left[\left(\frac{X_T - E_Q[X_T]}{\sqrt{\mathrm{Var}(X_T)}}\right)^N\right]$. Note that the mean of the log-asset return is adjusted beforehand by adding a mean correcting term in the stock price process in order to have a risk-neutral price process (compare with Equation (2.1)).

The VG model is fully described by three parameters that we calibrate by solving the system

$$\left\{\begin{aligned} v &= \mathrm{Var}^{\mathrm{Model}}\left(X_T\right) \\ s &= \mathrm{Skewness}^{\mathrm{Model}}\left(X_T\right) \\ k &= \mathrm{Kurtosis}^{\mathrm{Model}}\left(X_T\right) \end{aligned}\right. \tag{3.12}$$

if this system admits a solution which satisfies the domain conditions. For explicit expressions for the variance, skewness and kurtosis see Table 2.1. More precisely, the system (3.12) reduces (see Guillaume and Schoutens (2013a)) to

$$\left\{\begin{aligned} & c_1\sigma^6 + c_2\sigma^4 + c_3 = 0 \\ & \nu = \frac{s^2 v^3}{T^2\left(\sigma^2 + 2\frac{v}{T}\right)^2\left(\frac{v}{T} - \sigma^2\right)} \\ & \theta = \mathrm{sign}(s)\sqrt{\frac{\frac{v}{T} - \sigma^2}{\nu}} \end{aligned}\right. \tag{3.13}$$

where

$$\left\{\begin{aligned} c_1 &= \frac{\left(\frac{k}{3} - 1\right)T}{v s^2} \\ c_2 &= 3\frac{\frac{k}{3} - 1}{s^2} - 1 \\ c_3 &= 2\frac{v^2}{T^2}\left(1 - 2\frac{\frac{k}{3} - 1}{s^2}\right) \end{aligned}\right.$$

and where v, s and k denote the market-implied variance, skewness and kurtosis of the log-asset return for the time horizon T, respectively (see Equation (3.10)). The solution of Equation (3.13) can be directly obtained by finding the real root(s) of the cubic function $c_1x^3+c_2x^2+c_3=0$ where $x=\sigma^2$, which is given by Cardano's formula:

$$x=-\frac{c_2}{3c_1}+(R+\sqrt{D})^{1/3}+(R-\sqrt{D})^{1/3};$$

where

$$D=Q^3+R^2,\quad Q=-\frac{c_2^2}{9c_1^2}\quad\text{and}\quad R=-\frac{c_3}{2c_1}-\left(\frac{c_2}{3c_1}\right)^3.$$

The system (3.13) admits a solution $\{\sigma>0,\nu>0,\theta\in\mathbb{R}\}$ if and only if there exists at least one positive root of the cubic equation $c_1x^3+c_2x^2+c_3=0$, i.e. if and only if

$$6+3s^2-2k<0. \tag{3.14}$$

The existence domain of the VG moment calibration problem turns out to be independent on the value of the market implied variance v.

If the system (3.12) admits no solution, i.e. if the market-implied skewness and kurtosis do not satisfy the existence domain condition (3.14), we resort to solving the system (3.12), but we adjust beforehand the market-implied skewness and kurtosis by one of the two following methods (see Figure 3.9):

1. **Method 1**

   We replace (s,k) by the closest couple $(\tilde{s},\tilde{k}+\epsilon)$ belonging to the existence domain. In other words we minimize the distance between the adjusted couple skewness-kurtosis and the market implied one. This method is equivalent to allocating the same importance to the skewness and kurtosis matching in a standard calibration optimizer, and is termed the least squares moment calibration problem. The boundary of the existence domain under the exponential VG models under investigation can be parametrized in terms of a parabola as follows:

   $$k=a+bs^2,$$

   with $a=3$ and $b=1.5$. Hence, the closest couple on the boundary can be computed by minimizing the distance $\sqrt{\left(\tilde{k}-k\right)^2+(s-\tilde{s})^2}$, where $\tilde{k}=a+b\tilde{s}^2$. In other words, we have to solve the cubic equation $4b^2\tilde{s}^3+2(1+2b(a-k))\tilde{s}-2s=0$, where $\tilde{s}$ is the closest real root to the market-implied skewness s. To ensure that the adjusted couple skewness-kurtosis belongs to the existence domain whatever the value of $\tilde{s}$ (and so $\tilde{k}$), we shift the couple $(\tilde{s},\tilde{k})$ upwards by adding a small constant $\epsilon>0$ to the kurtosis.

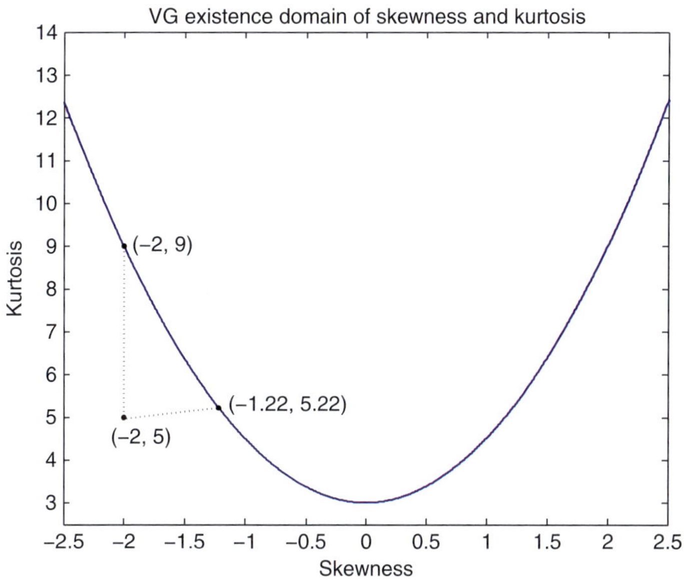

Figure 3.9 Calibration methods

2. **Method 2**
   We replace (s, k) by the closest couple $(\bar{s} = s, \bar{k} + \epsilon)$, i.e. we match the market-implied skewness and we adjust the kurtosis such that the adjusted couple skewness-kurtosis belongs to the existence domain. In other words, we set $\bar{s} = s$ and $\bar{k} = a + bs^2$. In the numerical example, we typically set ϵ equal to 0.1.

Example 3.6 *Let us assume we have estimated a skewness of -2 and a kurtosis of 5. Then method 1 looks for the point $(\tilde{s}, \tilde{k})$ on the parabola closest to our original point $(-2, 5)$. A simple calculation gives $(\tilde{s} = -1.22, \tilde{k} = 5.22)$. Method 2 just takes for $\bar{s} = s = -2$ and for $\bar{k} = 3 + 1.5s^2 = 9$. Both projections are shown in Figure 3.9.*

We finally remark that the above methods can also be applied for moments (or variance, skewness and kurtosis) obtained from a historical time-series. In Section 7.3, we will use this technique to fit a VG distribution on the basis of a historical data set.

3.3 Monte Carlo Simulation

Next, we look at possible simulation techniques for some processes encountered so far. A Lévy process can in general be simulated based on a compound Poisson approximation. However, typically, for very specific processes like the VG, other much faster techniques are available.

We assume we have random number generators at hand which can provide us Normal(0, 1) and Gamma(a, b) random numbers. Throughout, $\{v_n\}$ always denotes Normal(0, 1) random numbers; $\{g_n\}$ Gamma random numbers. A good reference book is Devroye (1986).

Recall that the risk-neutral price of a derivative is given by

$$\text{price} = \exp(-rT)E_Q[\text{payoff}],$$

where the expectation is taken under the pricing (risk-neutral) measure Q. Monte Carlo simulation can readily be applied to price European-type options: one simulates lots of stock price paths under the risk-neutral dynamics (hence employing the risk-neutral parameters) and evaluates for each path the payoff function. Assume we simulated N paths and that the payoff for path i ($i = 1, 2, \dots, N$) is denoted by payoff_i, then the price of the derivative is approximated by

$$\text{price} \approx \exp(-rT)\frac{\sum_{i=1}^{N} \text{payoff}_i}{N}.$$

Hence, under such a basic Monte Carlo simulation, the price is approximated by the discounted (equally weighted) average of the simulated payoffs.

In order to simulate the paths, we discretize time by taking time steps of size Δt, which we assume to be very small. It is not necessary to take equal time steps in a simulation, but for our purposes it is sufficient and convenient.

3.3.1 Brownian Motion

Recall that standard Brownian motion $W = \{W_t, t \geq 0\}$ has Normal distributed independent increments. One can simulate (by the Euler scheme) the value of the Brownian motion at the time points $\{n\Delta t, n = 0, 1, \dots\}$:

$$W_0 = 0, \quad W_{n\Delta t} = W_{(n-1)\Delta t} + \sqrt{\Delta t}\, v_n.$$

In Figure 3.10, 20 paths of a standard Brownian motion are shown.

3.3.2 The Gamma Process

Note that, when X is Gamma(a, b), then for $c > 0$, X/c is Gamma(a, bc). So we need only a good generator for Gamma$(a, 1)$ random numbers. Most mathematical software programs have built-in random gamma number generators.

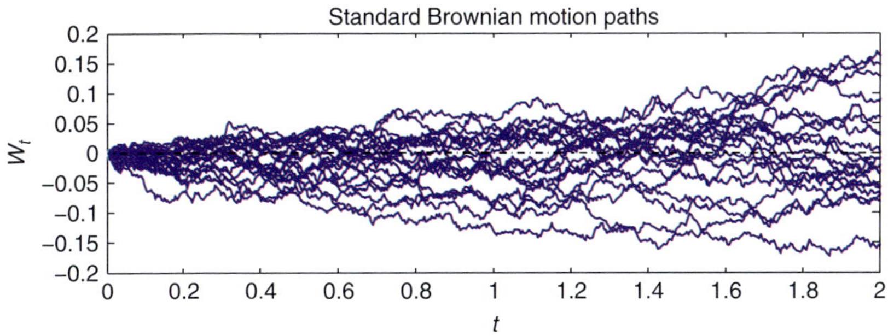

Figure 3.10 20 sample paths of a standard Brownian motion

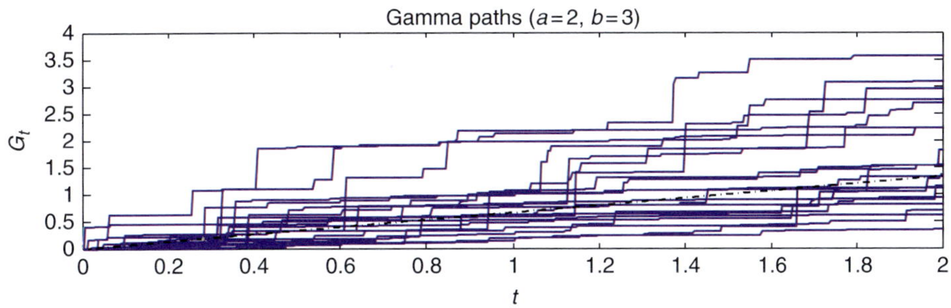

Figure 3.11 20 sample paths of a Gamma process ($a = 10$ and $b = 20$)

To simulate a sample path of a Gamma process $G = \{G_t, t \geq 0\}$, where G_t follows a Gamma(at, b) law at time points $\{n\Delta t, n = 0, 1, \ldots\}$:

- Generate independent Gamma($a\Delta t, b$) random numbers $\{g_n, n \geq 1\}$;
- Then set $G_0 = 0$ and

$$G_{n\Delta t} = G_{(n-1)\Delta t} + g_n, \quad n \geq 1.$$

In Figure 3.11, 20 sample paths of a Gamma process ($a = 10$ and $b = 20$) are shown.

3.3.3 The Variance Gamma Process

VG as the Difference of Two Gamma Processes

A VG process is the difference of two independent Gamma processes. More precisely, a VG process $X^{(VG)}$ with parameters $C, G, M > 0$ can be decomposed as

$$X_t^{(VG)} = G_t^{(1)} - G_t^{(2)},$$

where $G^{(1)}$ is a Gamma process with parameters $a = C$ and $b = M$, and $G^{(2)}$ is a Gamma process with parameters $a = C$ and $b = G$.

Similarly, VG(C, G, M)-random numbers h_n can be obtained as

$$h_n = g_n - \tilde{g}_n,$$

where g_n and $\tilde{g}_n$ are independent gamma variables with g_n Gamma(C, M) and $\tilde{g}_n$ Gamma(G, M) distributed.

To simulate a VG process $X^{(VG)}$, we first simulate VG$(\Delta t C, G, M)$-random numbers h_n and set

$$X_{n\Delta t} = X_{(n-1)\Delta t} + h_n, \quad n \geq 1,$$

with $X_0 = 0$.

VG as Time-Changed Brownian Motion

A VG process $X^{(VG)}$ with parameters (σ, ν, θ) can be seen as a subordination of a standard Brownian motion W with drift by a Gamma process G with parameters $a = 1/\nu$ and $b = 1/\nu$.

We have

$$X_t^{(VG)} = \theta G_t + \sigma W_{G_t}.$$

VG(σ, ν, θ)-random numbers h_n can be obtained out of Gamma(ν^{-1}, ν^{-1}) random numbers g_n and Normal(0, 1) numbers v_n:

$$h_n = \theta g_n + \sigma \sqrt{g_n} v_n.$$

Hence, to simulate a sample path of a VG process $X = \{X_t, t \geq 0\}$, where X_t follows a VG$(\sqrt{t}\sigma, \nu/t, \theta t)$ law at time points $\{n\Delta t, n = 0, 1, \ldots\}$:

- Generate independent VG$(\sqrt{\Delta t}\sigma, \nu/\Delta t, \theta \Delta t)$ random numbers h_n by simulating Gamma$(\Delta t \nu^{-1}, \nu^{-1})$ g_n and Normal(0, 1) numbers v_n and setting $h_n = \theta g_n + \sigma \sqrt{g_n} v_n$;
- Then set $X_0 = 0$ and

$$X_{n\Delta t} = X_{(n-1)\Delta t} + h_n, \quad n \geq 1.$$

In Figure 2.10, 20 VG sample paths ($\sigma = 0.2$, $\nu = 0.75$ and $\theta = -0.3$) are shown.

4

Conic Finance

4.1 Coherent Risk Measures

In this section we define *coherent measures of risks* (following Artzner et al., 1999). A set of properties of such coherent measures is given. The basic objects of the study are random variables X, also called *risks*, on the set of states of nature at a future date, say T. One way to interpret these random variables is as payoff of a derivative that needs to be paid out at time T.

So, we consider random variables X promised at T and defined on a probability space $(\Omega, \mathcal{F}, P)$. We assume $E[|X|] < \infty$. We call such an X a *risk*. A set of risks $\mathcal{A}$ is called *convex* if any convex combination of any two risk $X \in \mathcal{A}$ and $Y \in \mathcal{A}$ also belongs to the set $\mathcal{A}$: for any $0 \leq \alpha \leq 1$, $\alpha X + (1 - \alpha)Y \in \mathcal{A}$. Further, a set of risks is called a *cone* if, for each risk $X \in \mathcal{A}$ and any positive constant $c > 0$, cX also belongs to the set: $cX \in \mathcal{A}$.

In Figure 4.1 a cone of risks is visualized for the case when Ω consists of two points ω_1 and ω_2. Then any random variable X is represented as a point $(X(\omega_1), X(\omega_2))$ on the plane. The grey area (extended beyond the limits of the figure) constitutes a convex cone.

A risk measure is nothing but a functional that assigns a non-negative real number to a risk. Measuring risk is hence establishing a correspondence ρ between the space of random variables and the non-negative real numbers: $X \rightarrow \rho(X)$. Large values of $\rho(X)$ will tell us that X is very risky. If we think of X as a derivative's payoff, then potentially a very large payout needs to be made. If you promised X you should see $\rho(X)$ as the amount of cash (that you receive at time T) that should be added as a buffer so that the risk of paying out these potential large payoffs becomes "acceptable". The discounted amount of that cash is basically the price (paid at time zero) you will charge to take on board that risk. Alternatively, the random variable X may have some large negative values that are monies coming to you. One may then think of $\rho(X)$ as the amount of cash one may remove from

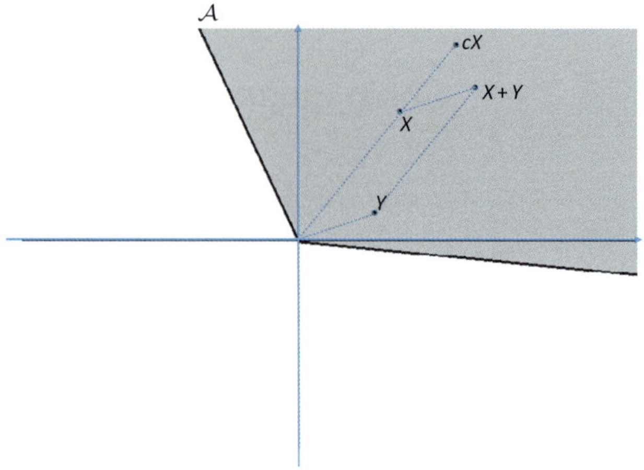

Figure 4.1 $\mathcal{A}$ (grey area) is a cone of risks

X and yet be "acceptable". In this formulation, $\rho(X)$ is seen as the equity capital in the position X as it is seen and discussed in financial circles.

Next, we will state four desirable properties of risk measures. If a risk measure satisfies all four of these properties we call it a *coherent risk measure*.

- Property 1 (Translativity): For all non-negative random variables X and each constant c, we have

$$\rho(X + c) = \rho(X) + c. \tag{4.1}$$

You can read this as the cash, $\rho(X+c)$, that needs to be added as a buffer so that $X+c$ becomes "acceptable", equals the cash $\rho(X)$, to make X acceptable, plus the cash c. As a consequence, one could say that cash is risk-free.

- Property 2 (Sub-additivity): For all non-negative random variables X and Y, we have

$$\rho(X + Y) \leq \rho(X) + \rho(Y). \tag{4.2}$$

You can read this as the cash, $\rho(X + Y)$, that needs to be added as a buffer so that sum $X + Y$ becomes "acceptable", is lower than the cash $\rho(X)$ to make X acceptable plus the cash $\rho(Y)$ to make Y acceptable. Sub-additivity reflects the idea that risk can be reduced by diversification.

- Property 3 (Positive homogeneity): For all non-negative random variables X and any positive constant $c > 0$, we have

$$\rho(cX) = c\rho(X). \tag{4.3}$$

You can read this as the cash, $\rho(cX)$, that needs to be added as a buffer so that cX becomes "acceptable", equals c times the cash $\rho(X)$ to make X acceptable. If you buy a certain number of units c in a risk, then the cash to make that position acceptable equals c times the cash to make one unit of X acceptable; taking positions into the same risk is not bringing in diversification.

- Property 4 (Monotonicity): For all non-negative random variables X and Y, we have

$$\text{if } P(X \leq Y) = 1, \text{ then } \rho(X) \leq \rho(Y). \tag{4.4}$$

You can read this as the cash, $\rho(X)$, that needs to be added as a buffer so that X becomes "acceptable", is always lower than the cash $\rho(Y)$ to make Y acceptable, if in all states of the world the risk (or the payoff to be made) X is smaller than or equal to the risk (or the payoff to be made) Y.

One can prove that a risk measure $\rho(X)$ of the form

$$\rho(X) = \sup_{Q \in \mathcal{M}} E_Q[X],$$

where $\mathcal{M}$ is a non-empty set of probability measures, satisfies the four above properties and hence is a coherent risk measure. Furthermore, Artzner et al. (1999) showed that each coherent risk measure (on a finite set of states of nature) is of this form (see also Delbaen (2002)).

4.2 Pricing on the Basis of Acceptable Risks

A modern financial economy is modelled on the recognition that all risks cannot be eliminated and risk exposures must be tolerated. Hence the set of acceptable risks must be defined as a financial primitive of the financial economy. We will define such economies in this section.

A zero-cost cash flow (at time T) should be seen as a linear combination of risks realizing at time T, which one enters at zero initial cost. The basic examples we will work with are a non-negative random variable minus a constant ($X - \exp(rT)c$) or a constant minus a non-negative random variable ($\exp(rT)c - X$). Next, we are going to specify which of these zero-cost cash flows the market is going to find acceptable as a primitive. Finding $X - \exp(rT)c$ acceptable is then seen as the market being willing to buy the risk X for the initial (time zero) price c; $\exp(rT)c - X$ is acceptable if the market is willing to sell the risk X at the initial price c.

First, let us consider making the set $\tilde{\mathcal{A}}$ of non-negative random variables acceptable at zero initial cost. Such variables should actually always be acceptable, since they are in fact arbitrages. You get a non-negative, potentially positive, cash flow for zero cost. Note that they constitute a convex cone as a set of random variables, being closed under convex combinations and scaling. Indeed, any convex combination of two non-negative random variables is again non-negative, and multiplying a non-negative random variable with a non-negative constant results again in a non-negative random variable. However, one could say that not only arbitrages are traded and that financial markets (should) accept zero-cost cash flows that are not necessary arbitrages.

In a second step, let us elaborate on the traditional risk-neutral valuation of risks. In that traditional one-price setting, the price or value, $V(X)$, of a financial derivative maturing at time T equals the discounted risk-neutral expectation of the payoff X:

$$V(X) = \exp(-rT)E_Q[X],$$

where Q is the risk-neutral measure. Assuming that the risk-neutral price is the "correct" price to trade on, the market will "accept" at initial zero cost the time T payoff given by

$$Z = X - \exp(rT)w, \quad \text{for } w \leq V(X).$$

Indeed, we have that

$$V(Z) = \exp(-rT)E_Q[X - \exp(rT)w] = V(X) - w \geq 0, \text{ for } w \leq V(X).$$

Hence the market is paying zero for something with non-negative risk-neutral value $V(Z)$. Equivalently, one can say that one is now actually selling X to the market for a price, w, below the risk-neutral market price $V(X)$: $V(X) \geq w$. Further, the market will also "accept" the time T payoff given by

$$Z' = \exp(rT)w - X, \quad \text{for } w \geq V(X).$$

Indeed, we have that

$$V(Z') = \exp(-rT)E_Q[\exp(rT)w - X] = w - V(X) \geq 0, \text{ for } w \geq V(X).$$

Hence the market is again paying zero for something with non-negative risk-neutral value $V(Z')$. One is now actually buying X from the market for a price, w, above the risk-neutral market price $V(X)$.

Therefore, let us consider the set of zero-cost cash flows defined as

$$\mathcal{A}^* = \left\{ Z | V(Z) = \exp(-rT)E_Q[Z] \geq 0 \right\}$$

as a potential candidate for the acceptable risks. Then clearly the set $\mathcal{A}^*$ contains the non-negative random variables $\tilde{\mathcal{A}}$. The set $\mathcal{A}^*$ is in fact an example of a largest possible convex cone containing the non-negative random variables as it is in fact a half-space.

Clearly, the set of acceptable zero-cost cash flows must contain the non-negative random variables; however, it is unlikely to be as generous and as large as a half-space $\mathcal{A}^*$. The half-space as a model for the set of acceptable zero-cost cash flows in a modern financial economy is just too generous. And indeed, in reality the market is not trading at the risk-neutral price. The prices at which one can (immediately) trade depend on the direction of the trade and are the current bid and ask prices.

We are then left with considering convex proper subsets of this half-space, but containing the non-negative random variables $\tilde{\mathcal{A}}$, as models for the set of potentially acceptable zero-cost cash flows by markets. The two-price economies (with bid and ask prices) we are setting up will maintain the conic structure of non-negative random variables by modelling acceptable zero-cost cash flows as a convex cone containing the non-negative random variables. However, the set of acceptable zero-cost cash flows $\mathcal{A}$ of a two-price economy will more precisely be a proper convex cone containing the non-negative random variables, in that it is not as large as a half-space: $\tilde{\mathcal{A}} \subset \mathcal{A} \subset \mathcal{A}^*$.

It is shown in Artzner et al. (1999) that any such cone of acceptable zero-cost cash flows can be defined by a convex set of probability measures $\mathcal{M}$ whereby

$$Z \in \mathcal{A} \iff \exp(-rT)E_Q[Z] \geq 0 \text{ for all } Q \in \mathcal{M}.$$

If the market now agrees to buy X for the price b or it agrees to sell X for the price a then

$$X - b\exp(rT) \in \mathcal{A} \text{ and } a\exp(rT) - X \in \mathcal{A}.$$

Equivalently for all $Q \in \mathcal{M}$:

$$\exp(-rT)E_Q[X - b\exp(rT)] = \exp(-rT)E_Q[X] - b \geq 0$$

and

$$\exp(-rT)E_Q[a\exp(rT) - X] = a - \exp(-rT)E_Q[X] \geq 0.$$

The best bid and ask prices for X provided by the market, denoted bid(X) and ask(X), respectively, are then given by

$$\text{bid}(X) = \exp(-rT) \inf_{Q \in \mathcal{M}} E_Q[X]; \tag{4.5}$$

$$\text{ask}(X) = \exp(-rT) \sup_{Q \in \mathcal{M}} E_Q[X] = \exp(-rT)\rho(X). \tag{4.6}$$

These are the two valuation operators for a two-price economy, and, by virtue of the infimum, the bid price functional is concave while the ask price functional is convex.

From the observation of equality between buying a random cash flow X and selling its negative $(-X)$, we deduce that the ask price for X is the negative of the bid price for $(-X)$, or that

$$\text{ask}(X) = -\text{bid}(-X). \tag{4.7}$$

In the classical market, with its law of one price, a zero-cost cash flow risk Z is acceptable if $\exp(-rT)E_Q[Z] \geq 0$, or in other words you pay zero to enter a position with a positive theoretical risk-neutral price. The set of acceptable risk is the half-space $\mathcal{A}^*$. Once we pass to the law of two prices we have just a convex cone containing the non-negative cash flows but no longer a half-space associated with one risk-neutral measure. However, we recognize that every convex set is also defined by the intersection of all the half-spaces that contain it, and this is also true for convex cones. Since our cone of acceptable cash flows contains the non-negative cash flows, we have a whole collection $\mathcal{M}$ of probability measures $Q \in \mathcal{M}$; with the property that a cash flow is acceptable to the market at zero cost only if its expectation is positive under **each** measure $Q \in \mathcal{M}$. Without loss of generality we may take the set $\mathcal{M}$ of measures to be a convex set as we may always include all the convex combinations. Every market is then defined by a convex cone of zero-cost cash flows acceptable to the market, and this cone has associated with it a convex set of probability measures $Q \in \mathcal{M}$ with acceptability equivalently defined as positive expectation under each $Q \in \mathcal{M}$.

We therefore refer to financial markets for the law of two prices as *conic*, given that they are defined by convex cones of acceptable cash flows.

4.3 No-Arbitrageable Markets

We may model different markets using different convex cones of acceptable zero-cost cash flows. For two markets we may have cones of acceptable zero-cost cash flows $\mathcal{A}_1$ and $\mathcal{A}_2$ with associated sets of measures $\mathcal{M}_1$ and $\mathcal{M}_2$. One may then wonder if the two markets may be arbitraged by buying some cash flow at the ask price from one market and selling it to the other market at its higher bid price. Now, for any cash flow X we determine the ask price $\text{ask}_i(X)$ in market $i = 1, 2$ as

$$\text{ask}_i(X) = \exp(-rT) \sup_{Q \in \mathcal{M}_i} E_Q[X], i = 1, 2.$$

Similarly one has that

$$\text{bid}_i(X) = \exp(-rT) \inf_{Q \in \mathcal{M}_i} E_Q[X], i = 1, 2.$$

Now, provided the set of supporting measures $\mathcal{M}_1$ and $\mathcal{M}_2$ have a common element, Q^* say, then

$$\text{ask}_1(X) \geq \exp(-rT)E_{Q^*}[X] \geq \text{bid}_2(X)\text{and } \text{ask}_2(X) \geq \exp(-rT)E_{Q^*}[X] \\ \geq \text{bid}_1(X).$$

Indeed, the ask price is always greater than the discounted expectation under Q^*, since the ask is the supremum of such a discounted expectation under each measure in respective measure sets $\mathcal{M}_1$ or $\mathcal{M}_2$. Similarly, the bid is always smaller than or equal to such a discounted expectation, since it is calculated as an infimum.

In conclusion, the bid price of market two is never above the ask price of market one, and the bid price of market one is never above the ask price of market two, if the set of supporting measures $\mathcal{M}_1$ and $\mathcal{M}_2$ has a common element.

4.4 Pricing Using Distorted Expectations

In the previous section we have shown how a market model may be constructed by specifying a set $\mathcal{M}$ of supporting measures. Bid and ask prices are then calculated by infimum and supremum of expectations over this set of supporting or test measures. Operational cones were defined by Cherny and Madan (2009) and employ concave distortion functions.

A concave distortion function is nothing more than a concave (distribution) function $\Psi(u)$ from the unit interval to itself:

$$\Psi : [0, 1] \to [0, 1] : u \mapsto \Psi(u).$$

This is, for example, shown in Figure 4.2. We give more, and more detailed, examples later.

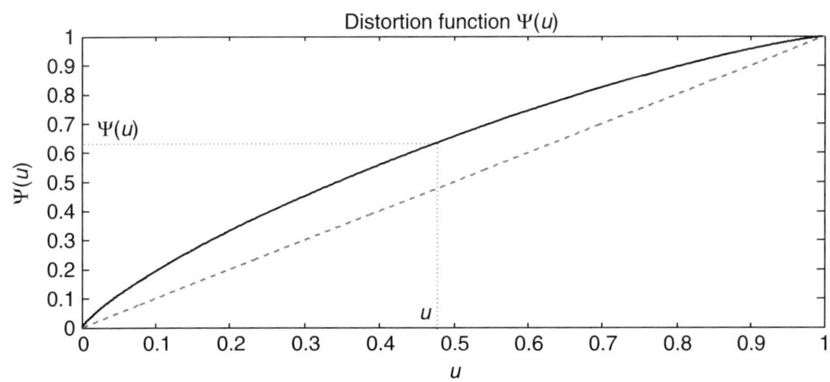

Figure 4.2 Example of a distortion function

If we restrict acceptability to be defined completely by the probability distribution of the associated risk, then the calculation of bid and ask prices can be made quite tractable using such concave distortion functions. Apart from letting acceptability, and hence the bid and ask price, depend *only* on the distribution function of the risk, we also need to ask for additivity for comonotone risks.

Two risks, X and Y, are said to be *comonotone* if they are actually completely driven by one single risk factor. In other words, there exists a random variable, U, on the unit interval such that

$$X = F_X^{[-1]}(U) \text{ and } Y = F_Y^{[-1]}(U),$$

where $F_X^{[-1]}(u)$ and $F_Y^{[-1]}(u)$ denotes the inverse of the distribution function of X and Y, respectively.

In general, from the sub-additivity property of coherent risk measures, we have that, for two random variables X and Y,

$$\begin{aligned}\text{bid}(X+Y) &\geq \text{bid}(X) + \text{bid}(Y)\\ \text{ask}(X+Y) &\leq \text{ask}(X) + \text{ask}(Y).\end{aligned}$$

These relationships can be seen as the effect of diversification. When X and Y are comonotone, one could say there is no scope for diversification, and for such a pair of random variables we in fact require that

$$\begin{aligned}\text{bid}(X+Y) &= \text{bid}(X) + \text{bid}(Y)\\ \text{ask}(X+Y) &= \text{ask}(X) + \text{ask}(Y).\end{aligned}$$

Under these two hypotheses of comonotone additivity and a dependence on just the distribution function, the results of Kusuoka (2001) imply that the bid and ask price must be an expectation under a concave distortion. More specifically, there must exist a concave distortion $\Psi(u)$ from the unit interval to itself such that, for any risk X with distribution function $F_X(x)$, its bid price is given by

$$\text{bid}(X) = \exp(-rT) \int_{-\infty}^{+\infty} x \mathrm{d}\Psi(F_X(x)). \tag{4.8}$$

We observe that for the bid price the lower quantiles are being re-weighted upwards while the higher quantiles are re-weighted downwards by the concavity of $\Psi(u)$. If $F_X(x)$ is the risk-neutral distribution of X, our bid price will hence be lower than the risk-neutral price, which is then nothing more than the above expectation but now without distortion:

$$\text{risk-neutral price}(X) = \exp(-rT) \int_{-\infty}^{+\infty} x \mathrm{d}F_X(x).$$

Its ask price is given by

$$\operatorname{ask}(X) = -\exp(-rT)\int_{-\infty}^{+\infty} x\mathrm{d}\Psi(F_{-X}(x)) \tag{4.9}$$

and now lower quantiles are being re-weighted downwards while the higher quantiles are re-weighted upwards. Our ask price will hence be higher than the risk-neutral price.

If X has a density function $f_X(x) = F'_X(x)$ and if also $\Psi(u)$ is differentiable, one has:

$$\operatorname{bid}(X) = \exp(-rT)\int_{-\infty}^{+\infty} x\Psi'(F_X(x))f_X(x)\mathrm{d}x. \tag{4.10}$$

Note that hence expectations under concave distortion can also be seen as expectations under the change of measure $\Psi'(F_X(x))$.

4.5 Conic Pricing as Choquet Expectations

Note that, using integration by parts, a traditional expectation of a random variable X may also be written as

$$E[X] = \int_{-\infty}^{+\infty} x\mathrm{d}F_X(x) = -\int_{-\infty}^{0} F_X(x)\mathrm{d}x + \int_{0}^{\infty} (1 - F_X(x))\,\mathrm{d}x.$$

One could say that the difficulty associated with the expectation operator is its uniform treatment of all probabilities. We are well aware that our experience with extreme tail events and their probabilities, i.e. $F_X(x)$ for very negative x (left tail) and $1 - F_X(x)$ for large x (right tail), rests on weak foundations and hence these probabilities are to be distrusted. A conservative approach would lift the probabilities on the left associated with losses and lower the probabilities on the right associated with gains.

Such a probability adjustment is accomplished by expectation with respect to a particular non-additive probability. Using a distortion given by a concave distribution function $\Psi(u)$, $0 \leq u \leq 1$, we define the distorted expectation or Choquet expectations, also termed non-linear expectation, as

$$\mathcal{D}^{\Psi}[X] = \int_{-\infty}^{+\infty} x\mathrm{d}\Psi(F_X(x)) = -\int_{-\infty}^{0} \Psi(F_X(x))\,\mathrm{d}x + \int_{0}^{\infty} (1 - \Psi(F_X(x)))\,\mathrm{d}x$$

that simultaneously raises the left tail and lowers the right tail. Importantly, the distorted expectation leverages the state ordering embedded by the variable X being evaluated. Distorted probabilities are not just functions of the state but also the order induced by the variable to be evaluated itself. These (non-linear) expectations, which are sub-additive or super-additive integrals, go back to the mathematician Gustave Choquet (see Choquet, 1953).

If X represents a cash flow, paid out at time T, then bid and ask prices may also be seen as expectations under a non-additive probability using Choquet expectations. More precisely, for the bid price one may apply integration by parts to deduce that

$$\begin{aligned} \text{bid}(X) &= \exp(-rT)\left(-\int_{-\infty}^{0} \Psi\left(F_X(x)\right)\mathrm{d}x + \int_{0}^{\infty}\left(1-\Psi\left(F_X(x)\right)\right)\mathrm{d}x\right) \\ &= \exp(-rT)\mathcal{D}^{\Psi}\left[X\right]. \end{aligned}$$

Now define the complementary distortion by

$$\widehat{\Psi}(u) = 1 - \Psi(1-u).$$

The above becomes somewhat more symmetrical:

$$\text{bid}(X) = \exp(-rT)\left(-\int_0^{\infty} \Psi(1-F_{-X}(x))\mathrm{d}x + \int_0^{\infty}\widehat{\Psi}(1-F_X(x))\mathrm{d}x\right).$$

The equivalent expression for the ask price is

$$\text{ask}(X) = \exp(-rT)\left(-\int_0^{\infty} \Psi(1-F_{X}(x))\mathrm{d}x + \int_0^{\infty}\widehat{\Psi}(1-F_{-X}(x))\mathrm{d}x\right).$$

Further observe that Ψ is both concave and bounded below by the identity function, while $\widehat{\Psi}$ is convex and bounded above by the identity. Therefore, the bid and ask price is a discounted expectation under a non-additive probability. One raises the probability of tails representing losses while lowering the probability of tails associated with gains.

These distortions are eminently reasonable, and call into question the use of undistorted expectations under classical additive probability. It is clear that our experience with extreme and rare tail events is limited, and justifies their distortion. Furthermore, we reiterate that events do not have tails, but outcomes do, and conservatism leads us to distort upwards the tails related to loss outcomes while we simultaneously distort downwards the tails for gain outcomes. As a consequence the expectations to be employed in decision making are rendered non-linear. Choquet expectations deliver the required non-linearity via the use of non-additive probabilities.

4.6 Indexes of Acceptability

The above concepts were actually introduced in the context of performance measures and indexes of acceptability of cash flows. A random cash flow will have a high acceptability level if its distribution function withstands high levels of stress or equivalently give rise to a stressed sampling that still has a positive expectation. The level of acceptability is then proportional to the level of stress. An illustrative example is provided by constructing the expectation of the minimum of n independent draws from a distribution of the cash flow to establish. By the positivity of such a stressed expectation, the cash flow is said to be acceptable at level n.

More precisely, we get that α is an acceptability index if and only if there exists an increasing one-parameter family of coherent risk measures $\{\rho_\lambda, \lambda \geq 0\}$ with the property that $\alpha(Z)$ is the largest level λ such that cash flow Z is acceptable to the level λ:

$$\alpha(Z) = \sup\{\lambda \geq 0 : Z \in \mathcal{A}_\lambda\}$$

where

$$Z \in \mathcal{A}_\lambda \iff \exp(-rT)E_Q[Z] \geq 0 \text{ for all } Q \in \mathcal{M}_\lambda,$$

with $\mathcal{M}_\lambda$ the convex set of probability measures associated with the coherent risk measure ρ_λ.

An example of such an increasing family of coherent risk measures is easily given using the tools introduced above. Let us, for example, consider a family of coherent risk measures arising from a particular distortion function $\Psi_\lambda(u)$ depending on a parameter λ, satisfying some properties.

Then, if

$$\rho_\lambda(X) = \int_{-\infty}^{+\infty} y \mathrm{d}\Psi_\lambda(F_X(y)), \tag{4.11}$$

we have that $\{\rho_\lambda, \lambda \geq 0\}$ is a family of risk measures and we can associate with it an acceptability index.

If one has a given bid or ask price and the distribution function of the related cash flow, one could calculate the particular λ that one needs to use in the given distortion to obtain the given price. We illustrate this later in detail in Section 6.1 after we have provided some practical useful examples of concrete distortion functions.

4.7 Distortion Functions

In the above subsection we employed distortion functions as an operational tool to calculate bid and ask prices. There are many possible distortion functions and some are more suited than others for our purposes. In this subsection we investigate a series of potential distortion function candidates.

4.7.1 The MINVAR Distortion Function

A very intuitive concave distortion function is given by the function

$$\Psi_{\lambda}^{MINVAR}(u) = 1 - (1-u)^{1+\lambda}, \qquad \lambda \geq 0.$$

One can show that, for an integer n,

$$\rho_n(X) = \int_{-\infty}^{+\infty} y \mathrm{d}\Psi_n(F_X(y)) = E[\min\{X_1, \ldots, X_{n+1}\}],$$

where $X_1, \ldots, X_{n+1}$ are independent draws of X, where we use the notation of Equation (4.11). We have that n is then the largest number such that the expectation of the minimum of $n+1$ draws from the cash flow distribution is still positive.

Indeed, the complement of the cdf of the minimum of $n+1$ independent and identically distributed random variable with distribution function $F_X(x)$ is given by

$$\begin{aligned} P(\min\{X_1, \ldots, X_{n+1}\} > x) &= P(X_1 > x, \ldots, X_{n+1} > x) \\ &= P(X_1 > x) \ldots P(X_{n+1} > x) \\ &= (1 - F_X(x))^{n+1}. \end{aligned}$$

Hence,

$$P(\min\{X_1, \ldots, X_{n+1}\} \leq x) = 1 - (1 - F_X(x))^{n+1} = \Psi_n^{MINVAR}(F_X(x)).$$

By working with real numbers (instead of integers) in the above distortion function, we allow for a real number of draws; it is analogous to the extension of the factorial function to the positive half-line as the Gamma function.

4.7.2 The MAXVAR Distortion Function

Another distortion function is

$$\Psi_{\lambda}^{MAXVAR}(u) = u^{\frac{1}{1+\lambda}}, \qquad \lambda \geq 0.$$

Now, one can show that, for an integer n, $\rho_n(X) = E[Y]$ where Y is a random variable with the property that the maximum of $n+1$ independent draws of Y has the same distribution as X. Indeed, we have that the distribution function of the maximum of $n+1$ independent and identically distributed random variable with distribution function $F_Y(x)$ is given by

$$\begin{aligned} P(\max\{Y_1, \ldots, Y_{n+1}\} \leq x) &= P(Y_1 \leq x, \ldots, Y_{n+1} \leq x) \\ &= P(Y_1 \leq x) \ldots P(Y_{n+1} \leq x) \\ &= F_Y(x)^{n+1}. \end{aligned}$$

Because we have that the maximum of $n+1$ independent draws of Y has the same distribution as X, we have that $F_Y(x)^{n+1} = F_X(x)$ and hence

$$F_Y(x) = F_X(x)^{\frac{1}{n+1}} = \Psi_n^{MAXVAR}(F_X(x)).$$

This class of distortion measures is known in the insurance literature as the *proportional hazard transform* (see Wang, 1995).

4.7.3 The MAXMINVAR Distortion Function

Combining the MINVAR and the MAXVAR distortion functions, we consider

$$\Psi_\lambda^{MAXMINVAR}(u) = \left(1 - (1-u)^{\lambda+1}\right)^{\frac{1}{1+\lambda}}, \qquad \lambda \geq 0.$$

Now, one can show that, for an integer n, $\rho_n(X) = E[Y]$ where Y is a random variable with the property that the maximum of $n+1$ independent draws of Y has the same distribution as the minimum of $n+1$ independent drawn from X:

$$\max\{Y_1, \ldots, Y_{n+1}\} =^{law} \min\{X_1, \ldots, X_{n+1}\}.$$

One could see this as if we construct the worst-case scenario first using a MINVAR perspective followed by a MAXVAR perspective.

4.7.4 The MINMAXVAR Distortion Function

Another way of combining MINVAR and MAXVAR is to consider

$$\Psi_\lambda^{MINMAXVAR}(u) = 1 - \left(1 - u^{\frac{1}{\lambda+1}}\right)^{1+\lambda}, \qquad \lambda \geq 0.$$

Now, one can show that for an integer n, $\rho_n(X) = E[Y]$ where Y is a random variable with the property that

$$Y =^{law} \min\{Z_1, \ldots, Z_{n+1}\} \text{ and } \max\{Z_1, \ldots, Z_{n+1}\} =^{law} X,$$

with $Z_1, \ldots, Z_{n+1}$ independent and identically distributed.

Now, we construct the worst-case scenario first using a MAXVAR perspective followed by a MINVAR perspective.

In Figure 4.3 the MINMAXVAR distortion function is shown for different λ values.

This distortion can actually also be generalized to a two-parameter distortion:

$$\Psi_{\lambda,\gamma}^{MINMAXVAR2}(u) = 1 - \left(1 - u^{\frac{1}{\lambda+1}}\right)^{1+\gamma}, \qquad \lambda, \gamma \geq 0.$$

We call this distortion MINMAXVAR2 for its two parameters.

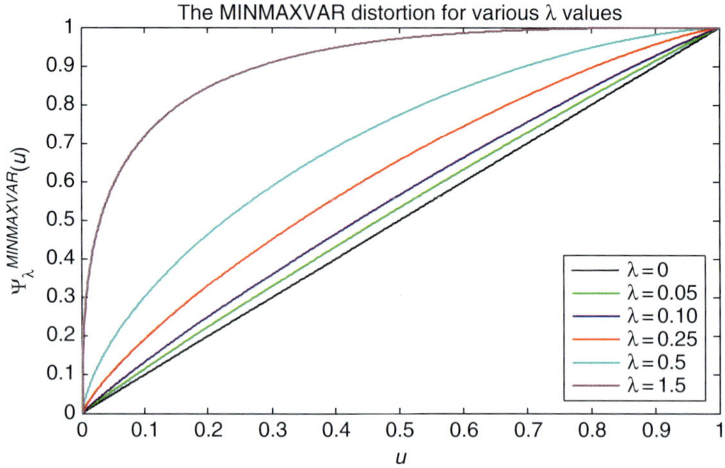

Figure 4.3 The MINMAXVAR distortion function for various λ values

4.7.5 The Wang Transform

The Wang transform (Wang, 2000, 2002a,b) is defined using the cumulative distribution function of the standard normal $\mathrm{N}(x)$ and its inverse function $\mathrm{N}^{[-1]}(u)$. It is defined for $0 < u < 1$ as follows:

$$\Psi_\lambda^{WANG}(u) = \mathrm{N}\left(\mathrm{N}^{[-1]}(u) + \lambda\right), \qquad \lambda \geq 0.$$

For $\lambda > 0$, this is a strictly concave function and a proper distortion function.

If we were to distort the cdf of a Normal(μ, σ^2) random variable we would get the cdf of the Normal$(\mu - \lambda\sigma, \sigma^2)$ variable:

$$\begin{aligned}\Psi_\lambda^{WANG}\left(\mathrm{N}\left(\frac{x-\mu}{\sigma}\right)\right) &= \mathrm{N}\left(\mathrm{N}^{[-1]}\left(\mathrm{N}\left(\frac{x-\mu}{\sigma}\right)\right) + \lambda\right)\\ &= \mathrm{N}\left(\frac{x-\mu}{\sigma} + \lambda\right)\\ &= \mathrm{N}\left(\frac{x-(\mu-\lambda\sigma)}{\sigma}\right).\end{aligned}$$

Similarly, if one were to distort the cdf of a lognormal(μ, σ^2) random variable, i.e. a random variable whose log has a Normal(μ, σ^2) distribution, we would get the cdf of a lognormal$(\mu - \lambda\sigma, \sigma^2)$ variable. Indeed, the cdf of a lognormal(μ, σ^2) random variable X is given by

$$P(X \leq x) = P(\log(X) \leq \log(x)) = \mathrm{N}\left(\frac{\log(x) - \mu}{\sigma}\right)$$

Table 4.1 *Limiting properties of distortion functions.*

Distortion	$\Psi'(u)$	$\lim_{u\to 0}\Psi'(u)$	$\lim_{u\to 1}\Psi'(u)$
MINVAR	$(1+\lambda)(1-u)^{\lambda}$	$(1+\lambda)$	0
MAXVAR	$u^{-\frac{\lambda}{1+\lambda}}/(1+\lambda)$	$+\infty$	$1/(1+\lambda)$
MAXMINVAR	$\left(1-(1-u)^{\lambda+1}\right)^{-\frac{\lambda}{\lambda+1}}(1-u)^{\lambda}$	$+\infty$	0
MINMAXVAR	$\left(1-u^{\frac{1}{\lambda+1}}\right)^{\lambda}u^{-\frac{\lambda}{\lambda+1}}$	$+\infty$	0
WANG	$\frac{n\left(\mathrm{N}^{[-1]}(u)+\lambda\right)}{n\left(\mathrm{N}^{[-1]}(u)\right)}$	$+\infty$	0

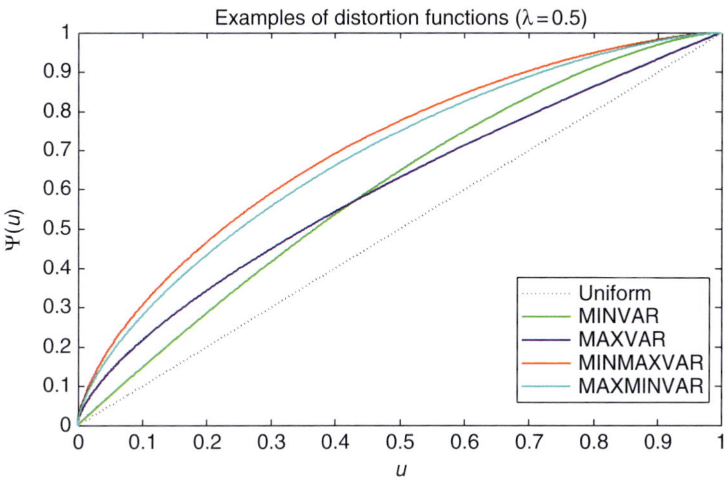

Figure 4.4 Examples of distortion functions

and therefore

$$\Psi_{\lambda}^{WANG}\left(\mathrm{N}\left(\frac{\log(x)-\mu}{\sigma}\right)\right)=\mathrm{N}\left(\mathrm{N}^{[-1]}\left(\mathrm{N}\left(\frac{\log(x)-\mu}{\sigma}\right)\right)+\lambda\right)$$
$$=\mathrm{N}\left(\frac{\log(x)-(\mu-\lambda\sigma)}{\sigma}\right).$$

4.8 Desired Properties of Distortion Functions

As is illustrated by Figure 4.4, there are clear differences between the above distortion functions, and one may ask which properties are desirable in this context of financial markets modelling.

First recall that expectations under concave distortions are also expectations under the measure change $\Psi'(F_X(x))$, as is shown in Equation (4.10). Hence, for x tending to negative infinity, and hence $F_X(x)$ tending to zero, one should induce loss aversion by ensuring that $\Psi'(u)$ tends to infinity as u tends to zero. Similarly, as x tends to positive infinity and $F_X(x)$ to unity, and hence u tends to unity, one should ensure against being enticed by large gains by requiring that $\Psi'(u)$ tends to zero. Distortions that have these properties are the MINMAXVAR and MAXMINVAR distortions and the Wang transform, as can be seen from Table 4.1, where $\lim_{u\to 0} \Psi'(u)$ and $\lim_{u\to 1} \Psi'(u)$ is given for all the distortions in the previous section.

5

Conic Pricing

In this chapter we first illustrate the conic pricing principle in a very simple setting: the binomial and trinomial tree model. We next illustrate bid and ask pricing in a general discrete setting. We further show how bid and ask pricing using distortion functions can easily be done in a Monte Carlo framework. Next, we move to the Black–Scholes setting and calculate the bid and ask prices of vanilla options in semi-closed form and derive conic Black–Scholes implied volatilities. Finally, we elaborate on bid and ask pricing of vanillas in a general setting with the VG model serving the examples.

5.1 Conic Trees

5.1.1 Conic Binomial Trees

Assume the very simple setting of a one-step binomial tree of Section 2.7.1. The asset is assumed for simplicity not to pay dividends and the current asset price is S_0. We have a risk-free interest rate r. Recall that the one-step binomial tree model allows the asset price, after a time-step of $\Delta t > 0$, to move to only two possible values, namely uS_0 and dS_0, representing respectively the up-state and down-state. We assume $0 \leq d < \exp(r\Delta t) < u$ and hence the model is arbitrage-free.

Consider now a derivative paying out $f_u \geq 0$ in the up-state and $f_d \geq 0$ in the down-state as in Figure 2.11. The current price f of this derivative can be obtained in the risk-neutral setting as the discounted expected payoff under the risk-neutral measure:

$$f = \exp(-r\Delta t)\,(pf_u + (1-p)f_d)\,,$$

where the risk-neutral probability to move to the up-state is given by

$$p = \frac{\exp(r\Delta t) - d}{u - d}.$$

In a conic setting, we now price using distorted expectations. Assume that $f_d < f_u$ and recall that, risk-neutrally, f_d is attained with probability $1-p$ and f_u

with probability p. The cdf of the payout is hence represented in Figure 5.1 (corresponding to values of the setting for the ATM call of Example 5.1). It has a jump of size $1 - p$ in the point f_d and then jumps to the unity level at the point f_u with a jump of size p. Hence now under the distortion f_d is attained with probability $\Psi(1 - p)$ and f_u with probability $1 - \Psi(1 - p)$. The distorted cdf is also given in the same Figure 5.1.

We therefore have that

$$\text{bid} = \exp(-r\Delta t)\left((1 - \Psi(1 - p))f_u + \Psi(1 - p)f_d\right).$$

The ask price is given as the negative of discounted distorted expectation of the negative cash flow. Since now $-f_u < -f_d$ the corresponding cdf is given by a function showing a jump of size p in $-f_u$ and a jump of size $1 - p$ to unity in $-f_d$, as shown in Figure 5.2, the distorted cdf has a jump of size $\Psi(p)$ in $-f_u$ and of size $1 - \Psi(p)$ in $-f_d$. The ask price is hence given by

$$\text{ask} = \exp(-r\Delta t)\left(\Psi(p)f_u + (1 - \Psi(p))f_d\right).$$

Assume now the other situation where $f_d > f_u$. Then the risk-neutral cdf first has a jump of size p in the point f_u and then jumps to the unity level at the point f_d with a jump of size $1 - p$. Under distortion f_u is attained with probability

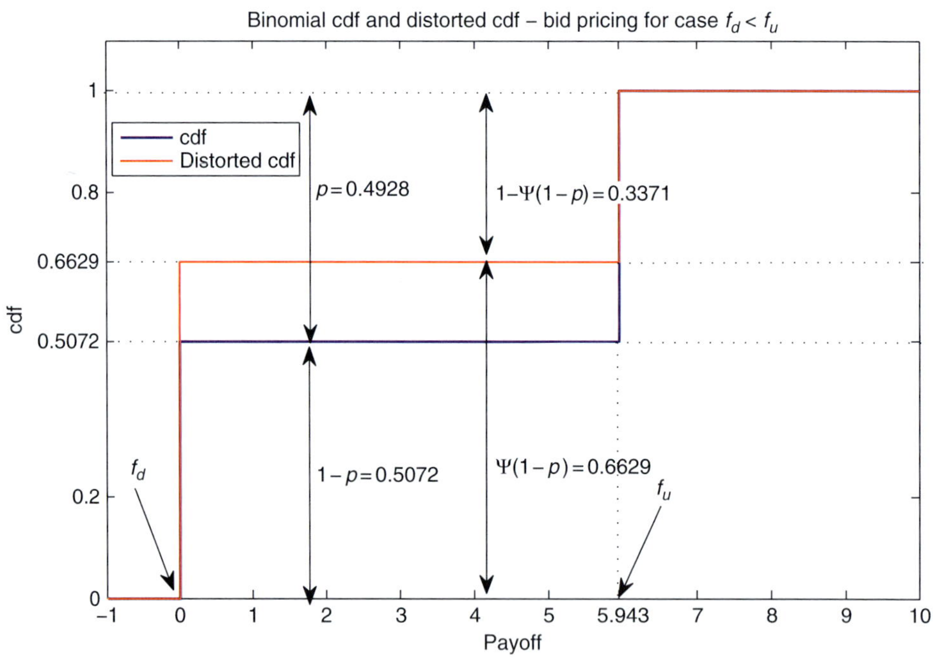

Figure 5.1 Binomial cdf and distorted cdf – bid pricing for case $f_d < f_u$

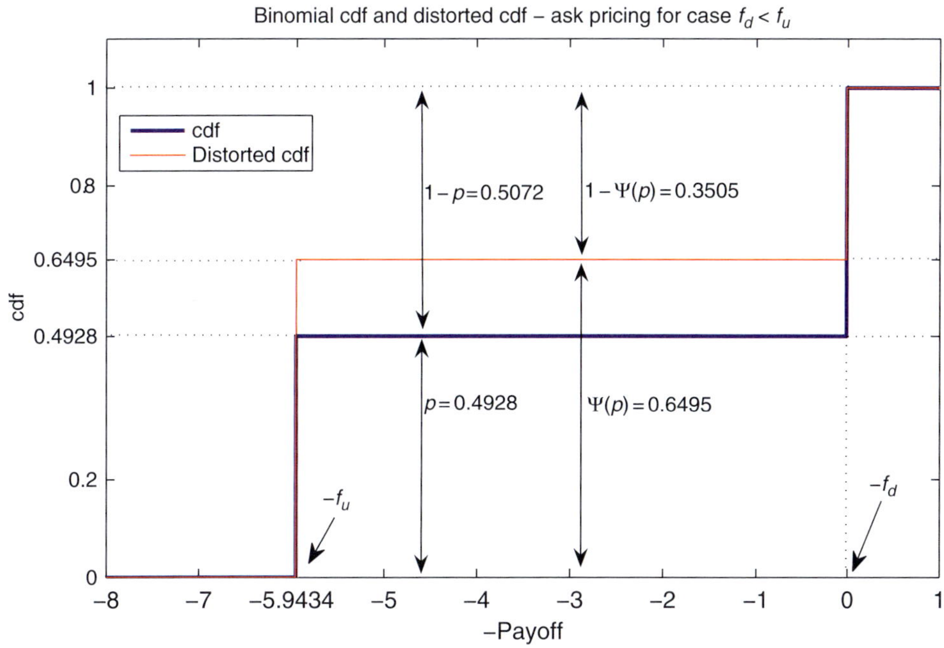

Figure 5.2 Binomial cdf and distorted cdf – ask pricing for case $f_d < f_u$

$\Psi(p)$ and f_d with probability $1 - \Psi(p)$. This is graphed in Figure 5.3 with values corresponding to the setting for the ATM put of Example 5.1. We therefore have that

$$\text{bid} = \exp(-r\Delta t)\left(\Psi(p)f_u + (1 - \Psi(p))f_d\right).$$

To determine the ask price, we note that now $-f_d < -f_u$ and the corresponding risk-neutral cdf is given by a function showing a jump of size $1 - p$ in $-f_d$ and a jump of size p to unity in f_u, as shown in Figure 5.4. The distorted cdf has a jump of size $\Psi(1 - p)$ in $-f_d$ and of size $1 - \Psi(1 - p)$ in $-f_u$ as is shown in the same Figure 5.4. The ask price is hence given by

$$\text{ask} = \exp(-r\Delta t)\left((1 - \Psi(1 - p))f_u + \Psi(1 - p)f_d\right).$$

We summarize the above in Table 5.1, where we provide the distorted probability to receive the up-state payoff. The distorted probability to receive the down-state payoff is then just one minus this probability. Note that

$$1 - \Psi(1 - p) \leq p \leq \Psi(p).$$

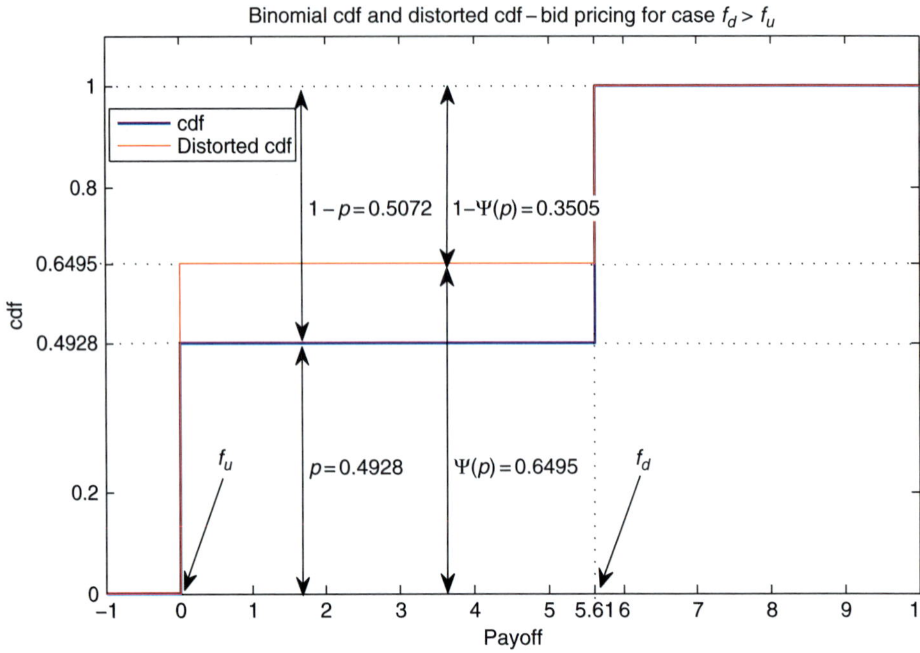

Figure 5.3 Binomial cdf and distorted cdf – bid pricing for case $f_d > f_u$

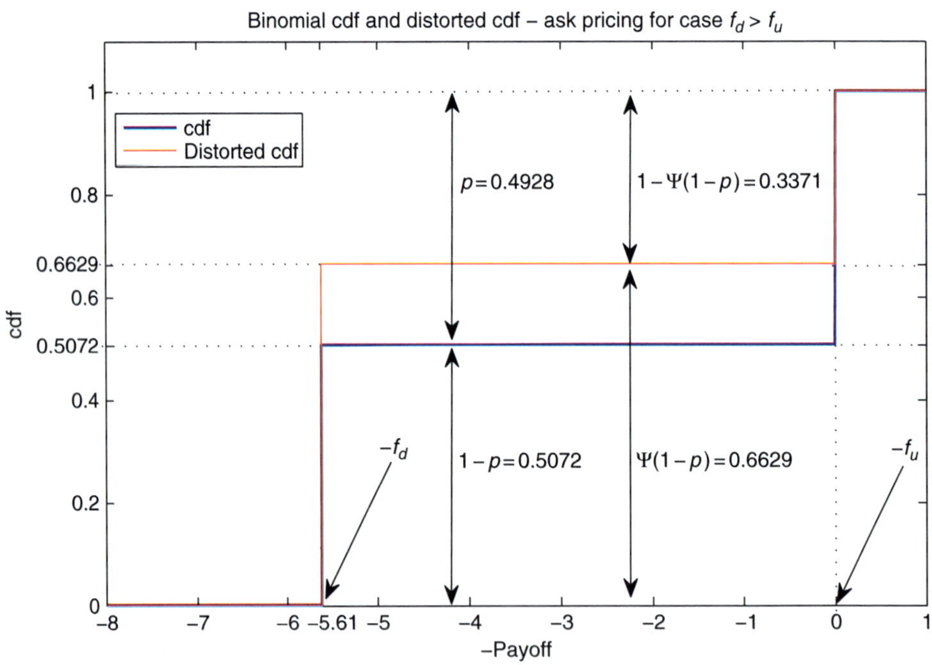

Figure 5.4 Binomial cdf and distorted cdf – ask pricing for case $f_d > f_u$

Table 5.1 *Risk-neutral and distorted up-state probabilities.*

	risk-neutral	bid	ask
$f_u > f_d$	p	$1 - \Psi(1-p)$	$\Psi(p)$
$f_d > f_u$	p	$\Psi(p)$	$1 - \Psi(1-p)$

Example 5.1 *Assume a non-dividend paying stock with a (yearly) volatility of 20% and a risks-free interest rate of 1%. We will price a one-month ATM call option. We will implement the so-called Cox–Ross–Rubinstein setting where*

$$u = \exp(\sigma\sqrt{\Delta t}) \text{ and } d = \exp(-\sigma\sqrt{\Delta t}).$$

With $\sigma = 0.20$ and $\Delta t = 1/12$, we have $u = \exp(\sigma\sqrt{\Delta t}) = 1.059434$ and $d = \exp(-\sigma\sqrt{\Delta t}) = 0.943900$. If $S_0 = 100$, we have that $f_u = 5.9434$ and $f_d = 0$. We will employ the MINMAXVAR distortion with parameter $\lambda = 0.25$. The risk-neutral probability to move up is $p = 0.4928$. Using the formulas in the above $f_u > f_d$ setting, we have a distorted probability to receive the up-state payoff of $1 - \Psi(1-p) = 0.3371$ to be used for the bid and $\Psi(p) = 0.6495$ to be used for the ask price calculation.

This leads to the following prices:

$$\begin{aligned}
\text{risk-neutral ATM call} &= \exp(-r\Delta t)\,(pf_u + (1-p)f_d) = 2.9264\\
\text{bid ATM call} &= \exp(-r\Delta t)\,((1-\Psi(1-p))f_u + \Psi(1-p)f_d) = 2.0021\\
\text{ask ATM call} &= \exp(-r\Delta t)\,(\Psi(p)f_u + (1-\Psi(p))f_d) = 3.8569.
\end{aligned}$$

For an ATM put, we have $f_u = 0$ and $f_d = 5.6100$. Since now $f_d > f_u$, we have a distorted probability to receive $f_u = 0$ of $\Psi(p) = 0.6495$ to be used for the bid and $1 - \Psi(1-p) = 0.3371$ to be used for the ask price calculation. One receives $f_d = 5.6100$ with a probability of one minus the latter.

This leads to the following prices:

$$\begin{aligned}
\text{risk-neutral ATM put} &= \exp(-r\Delta t)\,(pf_u + (1-p)f_d) = 2.8431\\
\text{bid ATM put} &= \exp(-r\Delta t)\,(\Psi(p)f_u + (1-\Psi(p))f_d) = 1.9648\\
\text{ask ATM put} &= \exp(-r\Delta t)\,((1-\Psi(1-p))f_u + \Psi(1-p)f_d) = 3.7156.
\end{aligned}$$

5.1.2 Geometrical Interpretation

Next, we will give a geometrical interpretation of acceptable risks and the bid and ask pricing in a binomial tree setting. This is illustrated in Figure 5.5 using the setting of Example 5.1.

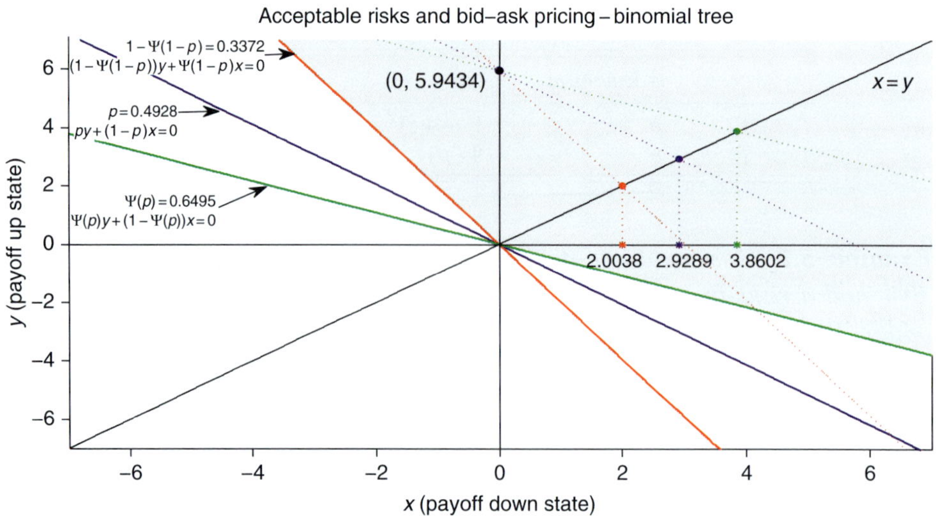

Figure 5.5 Acceptable risks and bid–ask pricing

A derivative or in general a random variable in such a binomial tree context is represented by a point (x_0, y_0) in the plane. The value x_0 is the payoff if we end in the down-state and y_0 is the payoff if we end in the up-state.

In a risk-neutral setting there is a (unique) risk-free probability p of ending in the up-state. In Figure 5.5 this probability is represented by the straight (blue) line through the origin with equation

$$py + (1 - p)x = 0,$$

and we call this the risk-neutral line. Actually, this line contains all the points, i.e. derivatives with payouts y in the up-state and payout x in the down-state, with a zero expectation under this risk-neutral measure. Since the risk-neutral price of such a risk is just the discounted value of this expectation, all the derivatives on this line have a zero initial risk-neutral price. Further, all the derivatives represented by points in the half-space above this line have a strictly positive price and all the derivatives represented by points in the half-space below the line have a strictly negative price.

Further, if we want to calculate the risk-neutral price of a particular derivative with a payout y_0 in the up-state and x_0 in the down-state, we proceed as follows. We first project the point (x_0, y_0) along the direction of our risk-neutral line on the diagonal with equation $x = y$. This projected point has coordinates solving the equations:

$$p(y - y_0) + (1 - p)(x - x_0) = 0 \text{ and } x = y.$$

The x- and y-coordinate of this projected point are the same (since the point lies on the diagonal) and are given by $py_0 + (1 - p)x_0$, which is the expected value of the payout under the risk-neutral probability. For the call option of Example 5.1, we get the value 2.9289, which, if we appropriately discount it, gives us the risk-neutral price of 2.9264.

In general, if we want to calculate the expected value of a payoff, we project the point representing this payoff on the diagonal line along the slope of the line representing the particular probability, p^* say, under which we want to calculate the expected value. Note that the slope of this line is given by $(p^* - 1)/p^*$.

In the two-price setting we work with distorted probabilities. Depending on the order of the payouts (either the up-state payoff is the largest or the down-state payoff is the largest), we have the distorted probabilities of Table 5.1. For the case, $f_u > f_d$, we employ for the bid price the distorted probability $1 - \Psi(1 - p)$ for an up-move and we employ the up-move probability $\Psi(p)$ for the ask price calculation. These probabilities are again represented by straight lines through the origin with equations respectively given by

$$(1 - \Psi(1 - p))y + \Psi(1 - p)x = 0 \text{ and } \Psi(p)y + (1 - \Psi(p))x = 0. \quad (5.1)$$

Bid and ask prices are calculated by projecting again the payoff on the diagonal along the direction of these lines. For our call, we get the value 2.0038 if we project along the direction of the former and 3.8602 if we use the direction of the latter. Discounting these values gives us a bid price of 2.0021 and an ask price of 3.8569.

If we had $f_u < f_d$, or in other words a point below the diagonal, the bid price would be obtained in a similar fashion but now by projecting along the direction of the line with equation $\Psi(p)y + (1 - \Psi(p))x = 0$, and the ask would be obtained after projection along the direction of the line $(1 - \Psi(1 - p))y + \Psi(1 - p)x = 0$.

With this in mind, we see that all the points above both lines will have a positive bid price (and hence also a positive ask price). This set of points is a cone, and this is represented by the grey area in Figure 5.5. The points on the border of this area have a zero bid price. All the points in the grey area are acceptable risk in the sense that, if you get them at zero cost, they are acceptable to an investor, since he gets the risk at zero cost, whereas a conservative value (the bid price) is actually positive.

Further, we can also see that the bid price is obtained as the infimum of the discounted expected values over a convex set of probability measures $\mathcal{M}$, and similarly the ask is obtained as the supremum of these discounted expectations over the same set $\mathcal{M}$. Observe first that, in this particular binomial tree setting, a probability measure is represented just by a single number: the probability of ending in the up-state. Here our set $\mathcal{M}$ is given by all the probability measures with an up-state probability in-between the probabilities $(1-\Psi(1-p))$ and $\Psi(p)$. This set is indeed a convex set, since we have that for two measures of this set with respective up-state probabilities p_1 and p_2 and for any $0 < \alpha < 1$, that $\alpha p_1 + (1-\alpha)p_2$ is a value also in-between $(1-\Psi(1-p))$ and $\Psi(p)$ and hence the measure with this up-state probability also belongs to $\mathcal{M}$. Further, we note that the risk-neutral probability is an element of this set $\mathcal{M}$. In our Figure 5.5, these probability measures would be represented by lines through the origin with slopes in-between the slopes of the lines in Equation (5.1). Expected values under such probability measures are obtained by projection along the slope of these lines on the diagonal and hence the bid price is the discounted value of the infimum of all these expectations and the ask price is the discounted value of the supremum. We hence see that the lines in Equation (5.1) are kind of the extremal or boundary lines. For a given payout, each of these two lines leads to an expected value, and the bid price is the lowest of the two expected values discounted and the ask price is the largest of these expected values discounted.

5.1.3 Conic Trinomial Trees

Next, we move to a trinomial tree setting as in Section 2.7.2. Now, there is an additional middle-state and the stock can now jump after a time step of $\Delta t > 0$ to three possible values, namely uS_0, mS_0 and dS_0, representing respectively the up-state, middle-state and down-state.

As mentioned in Section 2.7.2, a common choice for the factors u, m and d in terms of a given (yearly) volatility estimate σ is

$$\begin{aligned} u &= \exp((r-\sigma^2/2)\Delta t + \sigma\sqrt{3\Delta t}) \\ m &= \exp((r-\sigma^2/2)\Delta t) \\ d &= \exp((r-\sigma^2/2)\Delta t - \sigma\sqrt{3\Delta t}). \end{aligned}$$

A possible set of probabilities is then:

$$p_u = 1/6, \quad p_m = 2/3 \quad \text{and} \quad p_d = 1/6,$$

where p_u, p_m and p_d are respectively the probabilities of moving to the up-, middle- or down-state. Note that these probabilities are not exactly risk-neutral probabilities, but are very good approximations to a risk-neutral setting.

Consider now a derivative paying out f_u in the up-state, f_m in the middle-state and f_d in the down-state as in Figure 5.6.

The current price f of this derivative can be obtained in the risk-neutral setting as the discounted expected payoff under the risk-neutral measure.

$$f = \exp(-r\Delta t)\,(p_u f_u + p_m f_m + p_d f_d)\,.$$

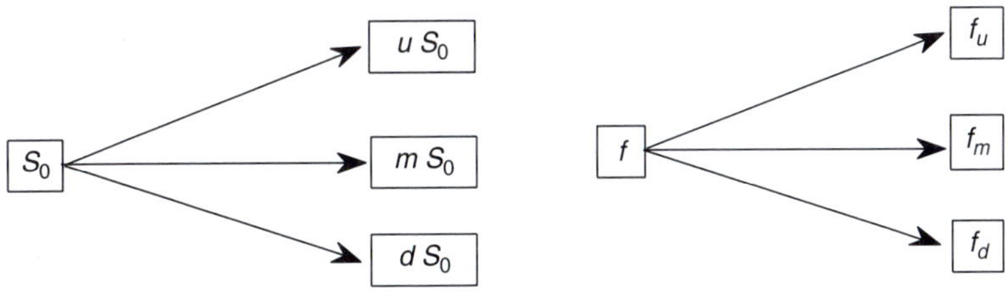

Figure 5.6 Trinomial tree

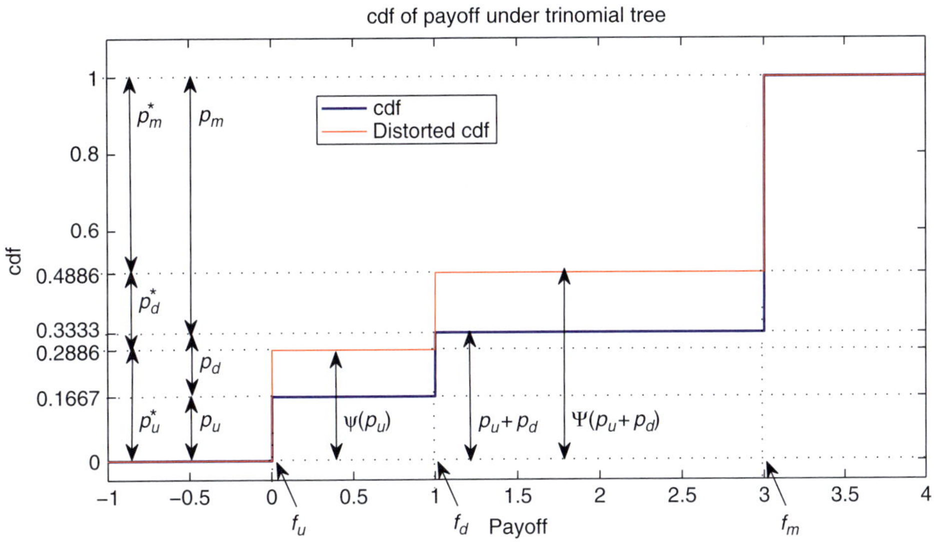

Figure 5.7 cdf and distorted cdf under a trinomial tree setting with $f_m \geq f_d \geq f_u$

To calculate the bid price of this derivative, we have to first sort the possible payoffs f_u, f_m and f_d. There are many possible combinations. We work out the case where $f_m \geq f_d \geq f_u$. Hence assume the lowest payoff is received in the up-state and the highest payoff in the middle-state.

To such a situation, a cdf as in Figure 5.7 is corresponding; the particular values correspond to the setting of Example 5.2. We also show a distorted cdf. One observes that the distorted probability to receive f_u is now $p_u^* = \Psi(p_u)$. With a probability $p_d^* = \Psi(p_u + p_d) - \Psi(p_u)$ one gets f_d and finally a payout f_m is received under distortion with a probability $p_m^* = 1 - \Psi(p_u + p_d)$. This leads to the bid price:

$$\text{bid} = \exp(-r\,\Delta t)\left(p_u^* f_u + p_m^* f_m + p_d^* f_d\right).$$

Example 5.2 *Consider a non-dividend-paying stock with* $S_0 = 100$ *and assume a (yearly) volatility of* 20% *and a risk-free interest rate of* 1%. *Let us price under* *the above trinomial setting a derivative that pays out zero in the up-state, one in the down-state and three in the middle-state:*

$$f_u = 0, \qquad f_m = 3 \qquad \textit{and} \qquad f_d = 1.$$

With $\sigma = 0.20$ *and* $\Delta t = 1/12$, *we have*

$$\begin{aligned} u &= \exp((r - \sigma^2/2)\Delta t + \sigma\sqrt{3\Delta t}) = 1.1043 \\ m &= \exp((r - \sigma^2/2)\Delta t) = 0.9992 \\ d &= \exp((r - \sigma^2/2)\Delta t - \sigma\sqrt{3\Delta t}) = 0.9041. \end{aligned}$$

We will employ the MINMAXVAR distortion with parameter $\lambda = 0.25$.

We start with the bid price calculation. The risk-neutral probability to move up is $p_u = 0.1667$. *We have a distorted probability to receive the up-state payoff* $p_u^* = \Psi(p_u) = 0.2886$. *Since* $p_u + p_d = 0.3333$, *we have that* $\Psi(p_u + p_d) = 0.4886$ *and hence that* $p_d^* = \Psi(p_u + p_d) - \Psi(p_u) = 0.2000$. *Therefore,* $p_m^* = 1 - p_u^* - p_d^* = 0.5114$.

For the ask price calculation we have to take the negative of the discounted distorted expectation of the negative payoff. Now $-f_m \leq -f_d \leq -f_u$, *and hence we start with first considering the probability of a jump to the middle state* $p_m = 0.6666$. *The distorted probability is now* $\tilde{p}_m = \Psi(p_m) = 0.7990$. *Further, we have that since* $p_m + p_d = 0.8333$, *we have that* $\Psi(p_m + p_d) = 0.9176$ *and hence that* $\tilde{p}_d = \Psi(p_m + p_d) - \Psi(p_m) = 0.1186$. *Therefore,* $\tilde{p}_u = 1 - \tilde{p}_m - \tilde{p}_d = 0.0824$.

This leads to the following prices:

$$\begin{aligned} \textit{risk-neutral} &= \exp(-r\Delta t)\,(p_u f_u + p_m f_m + p_d f_d) = 2.1649 \\ \textit{bid} &= \exp(-r\Delta t)\left(p_u^* f_u + p_m^* f_m + p_d^* f_d\right) = 1.7326 \\ \textit{ask} &= \exp(-r\Delta t)\,(\tilde{p}_u f_u + \tilde{p}_m f_m + \tilde{p}_d f_d) = 2.5136. \end{aligned}$$

5.2 Conic Discrete Pricing

The above binomial and trinomial models are examples of finite discrete market models. In general, such a finite discrete market model has a finite number of states, S_i, $i = 1, \ldots n$ at the maturity of interest: T, say. Assume the states are risk-neutrally attained with corresponding probabilities $0 < p_i < 1$, $i = 1, \ldots n$. We have $\sum_{i=1}^{n} p_i = 1$.

Next, we price a derivative (maturing at time T). Let us denote the payoff in the ith state (with stock price at S_i) by f_i. Then the risk-neutral price of the derivative is given by

$$f = \exp(-rT) \sum_{i=1}^{n} p_i f_i.$$

To calculate the bid and ask prices we have to distort the payoff's distribution function. For the bid price, we perform the following operations:

1. We sort the payoff vector $(f_i, i = 1, \dots, n)$ and we denote by $(f_{(i)}, i = 1, \dots, n)$ the sorted vector in increasing order, i.e.

$$f_{(1)} \leq f_{(2)} \leq \cdots \leq f_{(i)} \leq \cdots \leq f_{(n)}.$$

   We denote the corresponding probabilities by $p_{(i)}$. We emphasize that this is NOT to be confused with sorted probabilities. The probability $p_{(i)}$ is the probability that the payoff $f_{(i)}$ will be paid out. Note that $f = \exp(-rT) \sum_{i=1}^{n} p_i f_i = \exp(-rT) \sum_{i=1}^{n} p_{(i)} f_{(i)}$.
2. Next, we calculate the cumulative sum of the vector $(p_{(i)}, i = 1, \dots, n)$:

$$F_{(i)} = \sum_{j=1}^{i} p_{(j)}, i = 1, \dots, n.$$

   If all payoff values $f_{(i)}$ are different from each other, $F_{(i)}$ can be seen as a true cdf, i.e. as the probability one gets a payoff smaller or equal to $f_{(i)}$.
3. We now distort the vector $(F_{(i)}, i = 1, \dots, n)$ using our favourite distortion function Ψ, i.e. we calculate

$$(\Psi(F_{(i)}), i = 1, \dots, n).$$

   Note that $F_{(n)} = \Psi(F_{(n)}) = 1$.
4. From this vector, we calculate back individual distorted probabilities:

$$p^*_{(i)} = \Psi(F_{(i)}) - \Psi(F_{(i-1)}), i = 1, \dots, n,$$

   where $F_{(0)} = 0$.
5. Finally, the bid price is calculated as the discounted distorted expectation of the payoff:

$$f_{bid} = \exp(-rT) \sum_{i=1}^{n} p^*_{(i)} f_{(i)}.$$

To calculate the ask price, we perform similar operations. Indeed, the ask price is equal to the negative of the bid price of the derivative payout $-f_i$ in the ith state, $i = 1, \dots, n$.

1. Using the above notation, we now have

$$-f_{(n)} \leq -f_{(n-1)} \leq \cdots \leq -f_{(i)} \leq \cdots \leq -f_{(1)}.$$

2. We therefore now calculate the cumulative sum of the vector $(p_{(i)}, i = 1, \dots, n)$ backwards:

$$\tilde{F}_{(i)} = \sum_{j=n-i+1}^{n} p_{(j)}, i = 1, \dots, n.$$

3. We now distort this vector $(\tilde{F}_{(i)}, i = 1, \dots, n)$ using our favourite distortion function Ψ, i.e. we calculate

$$(\Psi(\tilde{F}_{(i)}), i = 1, \dots, n).$$

   Note that $\tilde{F}_{(n)} = \Psi(\tilde{F}_{(n)}) = 1$.
4. From this vector, we calculate now the individual distorted probabilities:

$$\tilde{p}_{(i)} = \Psi(\tilde{F}_{(i)}) - \Psi(\tilde{F}_{(i-1)}), i = 1, \dots, n,$$

   where $\tilde{F}_{(0)} = 0$.
5. Finally, the bid price is calculated as the discounted distorted expectation of the payoff:

$$f_{ask} = -\exp(-rT) \sum_{i=1}^{n} \tilde{p}_{(i)}(-f_{(i)}) = \exp(-rT) \sum_{i=1}^{n} \tilde{p}_{(i)} f_{(i)}.$$

5.3 Conic Bid and Ask Pricing of Exotic Options via Monte Carlo Simulations

As seen in Section 3.3, a Monte Carlo simulation can readily be used to price European options. The key idea is to simulate lots of stock price paths and evaluate for each path the payoff function. Assume we simulated N paths and that the payoff for path i ($i = 1, 2, \dots, N$) is denoted by payoff_i.

The price of the derivative in a risk-neutral setting is given by

$$\text{price} = \exp(-rT) E_Q[\text{payoff}]$$

and, using a Monte Carlo simulation, this price is approximated by

$$\text{price} \approx \exp(-rT) \frac{\sum_{i=1}^{N} \text{payoff}_i}{N}. \tag{5.2}$$

Hence, under such a basic Monte Carlo simulation, the price is the discounted weighted sum of the simulated payoffs with each payoff having an equal weight of $p = 1/N$. We are thus in a special case of the setting in the above Section 5.2.

Indeed, let us sort the payoffs:

$$\text{payoff}_{(1)} \leq \text{payoff}_{(2)} \leq \cdots \leq \text{payoff}_{(N)},$$

where now $\text{payoff}_{(1)}$ denotes the smallest payoff, $\text{payoff}_{(N)}$ the largest and in general $\text{payoff}_{(i)}$ is ranked at the ith place.

Note further that for the approximation the price

$$\text{price} \approx \exp(-rT)\frac{\sum_{i=1}^{N} \text{payoff}_i}{N} = \exp(-rT)\frac{\sum_{i=1}^{N} \text{payoff}_{(i)}}{N}.$$

In the conic setting, we attach higher weights to lower payoffs and lower weights to higher payoffs in order to calculate the bid price. Similarly, we attach higher weights to higher payoffs and lower weights to lower payoffs if we calculate the ask price. This distortion of the cumulative distribution function is the core of the conic pricing theory and can easily be calculated here.

Instead of assigning a weight $1/N$ to $\text{payoff}_{(i)}$, we will assign a distorted weight to it. Instead of having a weight $1/N$ attached to it, $\text{payoff}_{(i)}$ now receives a weight of

$$p_i^* = \Psi\left(\frac{i}{N}\right) - \Psi\left(\frac{i-1}{N}\right), \quad i = 1, \ldots, N,$$

if we are calculating the bid price. In Figure 5.8 the distorted weights are shown versus the uniform weights for a Monte Carlo exercise with 10,000 iterations and a MINMAXVAR distortion with $\lambda = 0.25$.

A similar reasoning shows that, for the ask price, we now assign a probability of

$$\tilde{p}_i = \Psi\left(\frac{N-i+1}{N}\right) - \Psi\left(\frac{N-i}{N}\right), \quad i = 1, \ldots, N,$$

to $\text{payoff}_{(i)}$.

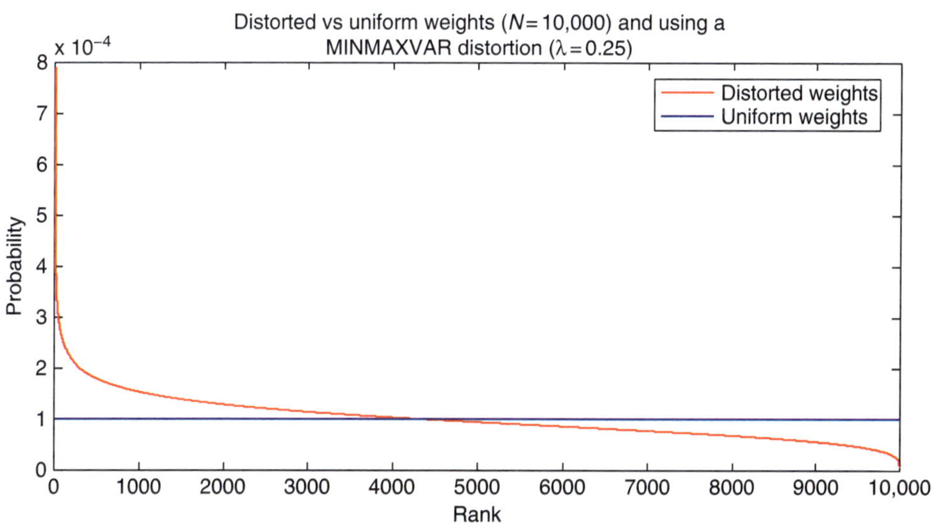

Figure 5.8 Distorted weights p^* under MINMAXVAR distortion with $\lambda = 0.25$ versus uniform weights for a Monte Carlo simulation with 10,000 simulations

In conclusion, we have:

$$\begin{aligned}\text{bid} &\approx \exp(-rT)\sum_{i=1}^{N} p_i^* \text{ payoff}_{(i)} \\ &= \exp(-rT)\sum_{i=1}^{N}\left(\Psi\left(\frac{i}{N}\right) - \Psi\left(\frac{i-1}{N}\right)\right)\text{payoff}_{(i)} \qquad (5.3)\end{aligned}$$

and

$$\begin{aligned}\text{ask} &\approx \exp(-rT)\sum_{i=1}^{N} \tilde{p}_i \text{ payoff}_{(i)} \qquad (5.4) \\ &= \exp(-rT)\sum_{i=1}^{N}\left(\Psi\left(\frac{N-i+1}{N}\right) - \Psi\left(\frac{N-i}{N}\right)\right)\text{payoff}_{(i)}. \qquad (5.5)\end{aligned}$$

Note that the new weights p_i^* and $\tilde{p}_i, i = 1, \dots, N$ can be precalculated and only depend on N (and the chosen distortion) but not on the paths simulated. Further applications and analysis can be found in Guillaume and Schoutens (2013b).

Example 5.3 *Let us price a European Call option on a non-dividend-paying stock S under a given model. We only do in this example $N = 10$ simulations, which is of course for practical purposes too low, but the only goal here is to numerically illustrate the above procedure. Assume the strike K equals 100, $T = 1$ and $r = 0.01$. We then show the final stock prices, $S_{T,i}, i = 1, \dots, N$ for the 10 simulations, the 10 payoffs, the sorted 10 payoffs in Table 5.2. The undistorted $(1/N)$ and distorted probabilities (p_i^* and $\tilde{p}_i$, $i = 1, \dots, N$), as well as the cdf $\bar{F}(x)$ and distorted cdf $\Psi(\bar{F}(x))$, are given in Table 5.3 where we have employed a MINMAXVAR distortion with parameter $\lambda = 0.25$.*

Note that we have a zero payoff four times. This is reflected in an undistorted cdf at zero of $\bar{F}(0) = 0.4$. This probability is distorted to a $\Psi^{MINMAXVAR}(\bar{F}(0), \lambda) = 0.5589$ value. This value corresponds to the sum of the first four distorted $\tilde{p}_i$, $i = 1, 2, 3, 4$ values: $0.5589 = 0.1940 + 0.1381 + 0.1196 + 0.1072$.

Applying Formulas 5.2, 5.3 and 5.5 leads to the following prices:

$$\begin{aligned}\textit{risk neutral price} &\approx 7.3858 \\ \textit{bid price} &\approx 4.5649 \\ \textit{ask price} &\approx 10.8108.\end{aligned}$$

Table 5.2 *Conic Monte Carlo pricing: payoffs.*

$S_{T,1}$	$S_{T,2}$	$S_{T,3}$	$S_{T,4}$	$S_{T,5}$	$S_{T,6}$	$S_{T,7}$	$S_{T,8}$	$S_{T,9}$	$S_{T,10}$
85.2	103.5	69.4	124.2	108.5	98.5	106.2	125.3	98.4	106.9
payoff_1	payoff_2	payoff_3	payoff_4	payoff_5	payoff_6	payoff_7	payoff_8	payoff_9	payoff_{10}
0	3.5	0	24.2	8.5	0	6.2	25.3	0	6.9
$\text{payoff}_{(1)}$	$\text{payoff}_{(2)}$	$\text{payoff}_{(3)}$	$\text{payoff}_{(4)}$	$\text{payoff}_{(5)}$	$\text{payoff}_{(6)}$	$\text{payoff}_{(7)}$	$\text{payoff}_{(8)}$	$\text{payoff}_{(9)}$	$\text{payoff}_{(10)}$
0	0	0	0	3.5	6.2	6.9	8.5	24.2	25.3

Table 5.3 *Conic Monte Carlo pricing: undistorted and distorted probabilities.*

x	$\text{payoff}_{(1)}$	$\text{payoff}_{(2)}$	$\text{payoff}_{(3)}$	$\text{payoff}_{(4)}$	$\text{payoff}_{(5)}$	$\text{payoff}_{(6)}$	$\text{payoff}_{(7)}$	$\text{payoff}_{(8)}$	$\text{payoff}_{(9)}$	$\text{payoff}_{(10)}$
$\bar{F}(x)$	0.4	0.4	0.4	0.4	0.5	0.6	0.7	0.8	0.9	1
$\Psi(\bar{F}(x))$	0.5589	0.5589	0.5589	0.5589	0.6562	0.7447	0.8248	0.8960	0.9569	1.0000
p	0.1	0.1	0.1	0.1	0.1	0.1	0.1	0.1	0.1	0.1
$\tilde{p}_i$	0.1940	0.1381	0.1196	0.1072	0.0973	0.0885	0.0801	0.0713	0.0609	0.0431
$\bar{p}_i$	0.0431	0.0609	0.0713	0.0801	0.0885	0.0973	0.107	0.1196	0.1381	0.1940

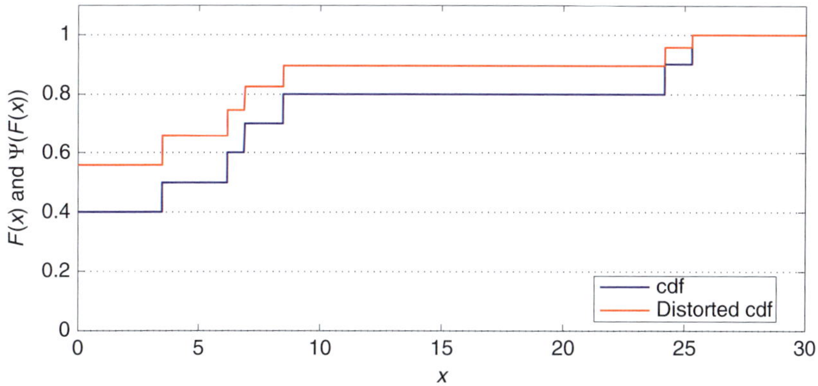

Figure 5.9 Distorted cdf under MINMAXVAR distortion with $\lambda = 0.25$

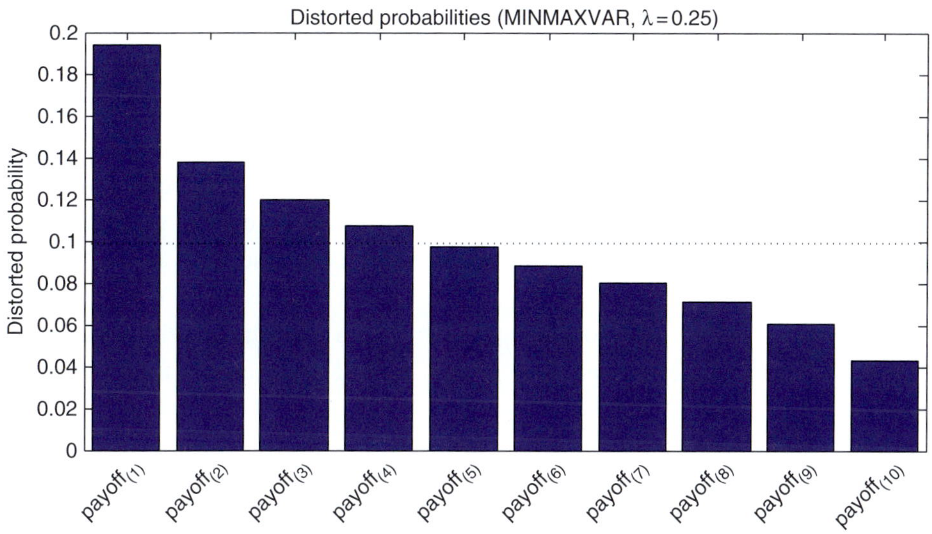

Figure 5.10 Distorted probabilities p^* under MINMAXVAR distortion with $\lambda = 0.25$

5.4 Conic Black–Scholes Vanilla Pricing

5.4.1 The European Call

Let us consider a European Call option priced under the Black–Scholes setting. Then the price of the European Call is given by

$$EC(K, T) = \exp(-qT)S_0\mathrm{N}(d_1) - K\exp(-rT)\mathrm{N}(d_2),$$

with

$$d_1(K) = \frac{\log(S_0/K) + (r - q + \sigma^2/2)}{\sigma\sqrt{T}} \text{ and } d_2(K) = \frac{\log(S_0/K) + (r - q - \sigma^2/2)}{\sigma\sqrt{T}}, \tag{5.6}$$

where we have made the dependency on the strike explicit in the notation.

The risk under investigation is the payoff of the call: $(S_T - K)^+$ and a simple calculation shows that

$$\begin{aligned} F_{(S_T-K)^+}(x) &= P((S_T - K)^+ \leq x) \\ &= 1 - P((S_T - K)^+ > x) \\ &= 1 - P(S_T > K + x) \\ &= 1 - \mathrm{N}(d_2(K + x)) \\ &= 1 - \mathrm{N}\left(\frac{\log(S_0/(K + x)) + (r - q - \sigma^2/2)}{\sigma\sqrt{T}}\right), \quad x \geq 0. \end{aligned}$$

This cdf is shown in Figure 5.11 for the setting $S_0 = 100$, $K = 85$, $r = 1\%$, $T = 1$, $\sigma = 0.20$ and $q = 0$. In this Figure the distorted cdf under a MINMAX-VAR distortion with parameter $\lambda = 0.1$ is also given. This distortion function is shown in Figure 5.12. For example, we see how the cdf at 10, or equivalently the probability of receiving a payoff smaller than or equal to 10 equals, without distortion, $F_{(S_T-K)^+}(10) = 0.4182$ and after distortion has a value $\Psi(F_{(S_T-K)^+}(10)) = \Psi(0.4182) = 0.4847$.

In Figure 5.13, the corresponding pdfs are shown as well as their average values of 17.99 and 15.21 respectively. Since the bid price equals the discounted

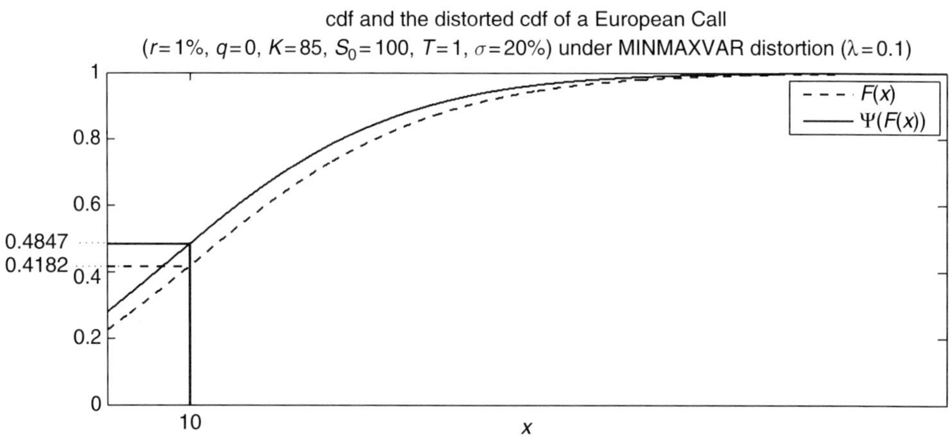

Figure 5.11 cdf and distorted cdf of a European Call

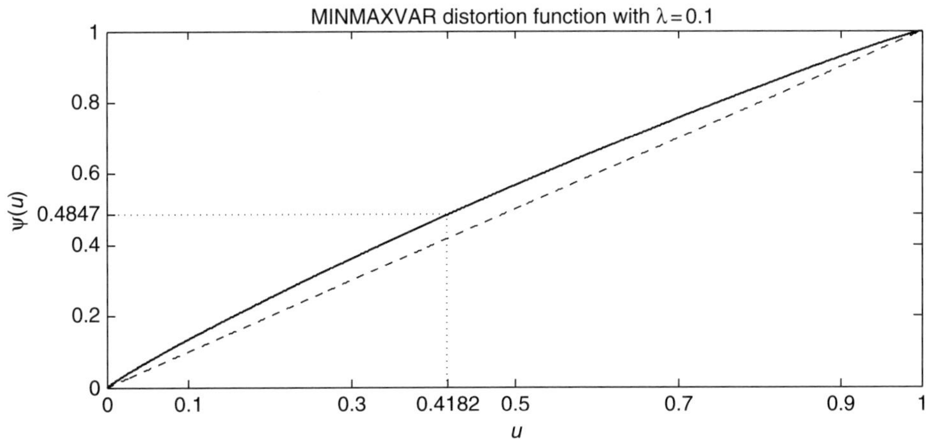

Figure 5.12 MINMAXVAR distortion function with $\lambda = 0.1$

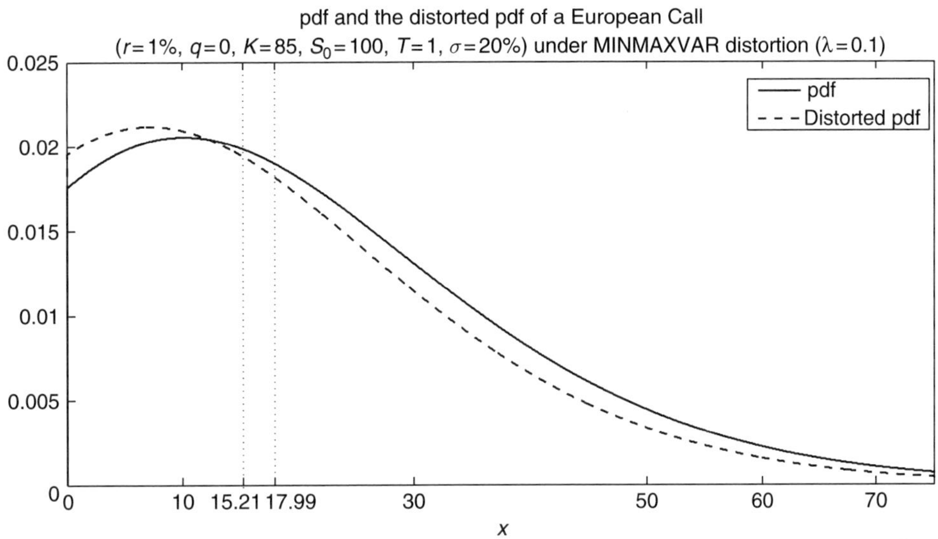

Figure 5.13 pdf and distorted pdf of a European Call

average value of the distorted payoff distribution, we can now easily numerically calculate

$$\text{bidEC}(K, T) = \exp(-rT) \int_0^{+\infty} x \mathrm{d}\Psi_\lambda(F_{(S_T-K)^+}(x)). \tag{5.7}$$

In our example of Figures 5.11 and 5.13, the risk-neutral price equals 17.81 ($= \exp(-rT)17.99$) and the bid price equals 15.06 ($= \exp(-rT)15.21$) under a MINMAXVAR distortion with parameter $\lambda = 0.1$.

For the ask price, we have to take the negative of the discounted mean of the distorted distribution of the negative payoff. We have

$$
\begin{aligned}
F_{-(S_T-K)^+}(x) &= P(-(S_T - K)^+ \leq x) \\
&= P((S_T - K)^+ \geq -x) \\
&= P(S_T > K - x) \\
&= \mathrm{N}(d_2(K - x)) \\
&= \mathrm{N}\left(\frac{\log(S_0/(K-x)) + (r - q - \sigma^2/2)}{\sigma\sqrt{T}}\right), \quad x \leq 0.
\end{aligned}
$$

Therefore one can calculate the ask as

$$
\mathrm{askEC}(K, T) = -\exp(-rT)\int_{-\infty}^{0} x\mathrm{d}\Psi_\lambda(F_{-(S_T-K)^+}(x)). \tag{5.8}
$$

5.4.2 The European Put

Next, we focus on the European Put. Under the Black–Scholes setting, the risk-neutral price of such a put is given by

$$
EP(K, T) = -\exp(-qT)S_0\mathrm{N}(-d_1) + K\exp(-rT)\mathrm{N}(-d_2),
$$

with d_1 and d_2 as above in Equation (5.6).

The risk under investigation is now the payoff of the put: $(K - S_T)^+$. A simple calculation shows that

$$
\begin{aligned}
F_{(K-S_T)^+}(x) = P((K - S_T)^+ \leq x) &= P(S_T \geq K - x) \\
&= \mathrm{N}(d_2(K - x)) \\
&= \mathrm{N}\left(\frac{\log(S_0/(K-x)) + (r - q - \sigma^2/2)}{\sigma\sqrt{T}}\right), \\
&\quad 0 \leq x \leq K
\end{aligned}
$$

and one can numerically calculate the bid as

$$
\mathrm{bidEP}(K, T) = \exp(-rT)\int_{0}^{K} x\mathrm{d}\Psi_\lambda(F_{(K-S_T)^+}(x)). \tag{5.9}
$$

For the ask price, we have to take the negative of the discounted average value of the distorted distribution of the negative payoff, and therefore we consider:

$$
\begin{aligned}
F_{-(K-S_T)^+}(x) &= P(-(K - S_T)^+ \leq x) \\
&= P((K - S_T)^+ \geq -x) \\
&= 1 - P(S_T > K + x)
\end{aligned}
$$

$$= 1 - \mathrm{N}(d_2(K + x))$$
$$= 1 - \mathrm{N}\left(\frac{\log(S_0/(K + x)) + (r - q - \sigma^2/2)}{\sigma\sqrt{T}}\right), \quad -K \leq x \leq 0,$$

and we get

$$\mathrm{askEP}(K, T) = -\exp(-rT)\int_{-K}^{0} x \mathrm{d}\Psi_\lambda(F_{-(K-S_T)^+}(x)). \tag{5.10}$$

Example 5.4 *As an example, we price a put with maturity one year and strike 90, while* $S_0 = 100$, $r = 0.02$, $q = 0.01$ *and* $\sigma = 0.2$. *We use the MINMAX-VAR distortion with parameter* $\lambda = 0.1$. *In Figure 5.14, the cdf* $F_{(K-S_T)^+}(x)$ *and distorted cdf* $\Psi_\lambda(F_{(K-S_T)^+}(x))$ *of the put payoff is shown. We observe that the probability to get a zero payoff equals* $F_{(K-S_T)^+}(0) = 0.6832$ *under no distortion and* $\Psi_\lambda(F_{(K-S_T)^+}(0)) = 0.7412$ *after distortion. Similarly, in Figure 5.15, we graph the cdf and the distorted cdf of the negative of the put payoff. On this figure one observes that the probability of getting a non-zero negative payoff equals* $F_{-(K-S_T)^+}(0) = 0.3168$ *under no distortion and* $\Psi_\lambda(F_{-(K-S_T)^+}(0)) = 0.3791$ *after distortion.*

Note that $F_{(K-S_T)^+}(0) + F_{-(K-S_T)^+}(0) = 1$, *but we have, in case of distortion,* $\Psi_\lambda(F_{(K-S_T)^+}(0)) + \Psi_\lambda(F_{-(K-S_T)^+}(0)) \neq 1$.

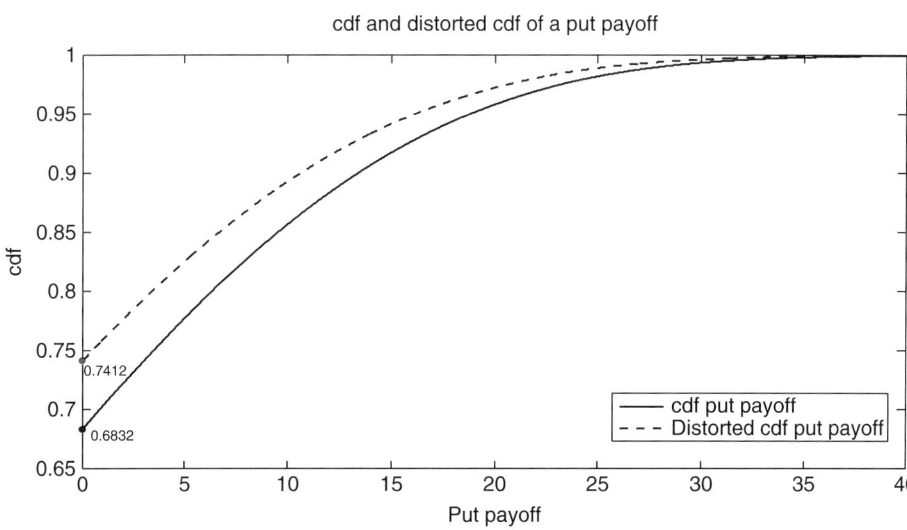

Figure 5.14 cdf and distorted cdf of a 90-strike put payoff under MINMAXVAR ($\lambda = 0.1$) under the Black–Scholes setting with $S_0 = 100$, $r = 0.02$, $q = 0.01$, $T = 1$ and $\sigma = 0.20$

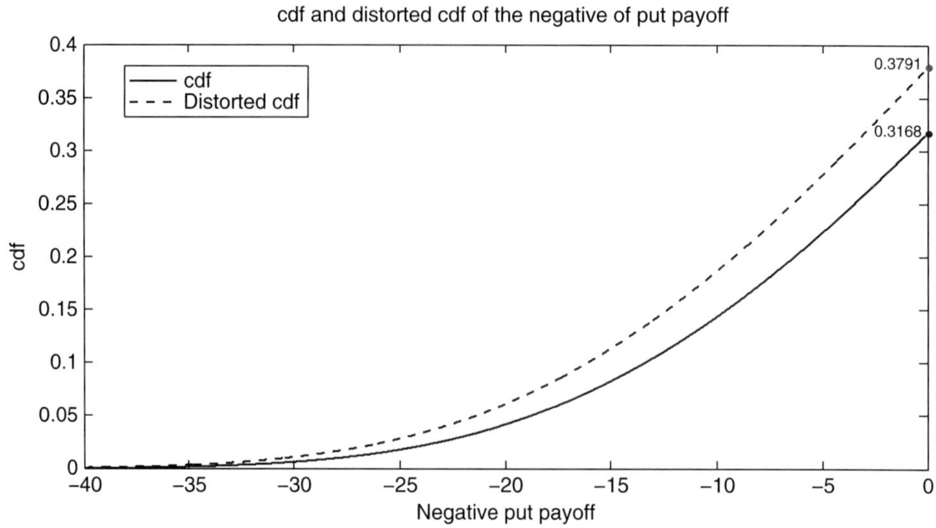

Figure 5.15 cdf and distorted cdf of the negative of a 90-strike put payoff under MINMAXVAR ($\lambda = 0.1$) under the Black–Scholes setting with $S_0 = 100$, $r = 0.02$, $q = 0.01$, $T = 1$ and $\sigma = 0.20$

If we perform a simple numerical integration and calculate the bid and ask price, we obtain 2.47 for the bid and 4.22 for the ask, while the risk-neutral price equals 3.26.

Example 5.5 *In case we employ the Wang transform of Section 4.7.5 as a distortion, we obtain a closed-from expression for the bid and ask prices of European Calls and Puts. Indeed, recall that if one were to distort the cdf of a log-normal(μ, σ^2) random variable, i.e. a random variable whose log has a Normal(μ, σ^2) distribution, we would get the cdf of a log-normal($\mu - \lambda\sigma, \sigma^2$) variable.*

Since, under the Black–Scholes setting, the distribution of S_T is log-normal and more precisely log-normal($\log S_0 + (r - q - \sigma^2/2)T, \sigma^2 T$) distributed, we have that

$$\Psi_\lambda^{WANG}(F_S(x)) = N\left(\frac{\log(x) - (\log S_0 + (r - q - \sigma^2/2)T - \lambda\sigma\sqrt{T})}{\sigma\sqrt{T}}\right),$$

which is the cdf of a log-normal($\log S_0 + (r - (q + \lambda\sigma/\sqrt{T}) - \sigma^2/2)T, \sigma^2 T$) distributed random variable. Therefore we have that

$$\begin{aligned} bidEC(K, T; S_0, r, q, \sigma,) &= \exp(-rT)\int_0^{+\infty} x d\Psi_\lambda^{WANG}(F_S(K + x; S_0, r, q, T, \sigma)) \\ &= \exp(-rT)\int_K^{+\infty} (x - K) d\Psi_\lambda^{WANG}(F_S(x; S_0, r, q, T, \sigma) \end{aligned}$$

$$= \exp(-rT) \int_K^{+\infty} (x - K) dF_S(x; S_0, r, q + \lambda\sigma/\sqrt{T}, T, \sigma)$$
$$= EC(K, T; S_0, r, q + \lambda\sigma/\sqrt{T}, \sigma),$$

where we have made the notation depend on the inputs. In conclusion, the

$$bidEC(K, T; S_0, r, q, \sigma,) = EC(K, T; S_0, r, q + \lambda\sigma/\sqrt{T}, \sigma)$$

and hence the bid price is just a traditional Black–Scholes price under a different dividend yield. More precisely, instead of using the dividend yield q, we now use in our calculations of the call's bid price a dividend yield $q + \lambda\sigma/\sqrt{T}$. Recognizing that a call option's price decreases if the dividend yield increases, we clearly see that the bid prices is lower than the risk-neutral price. Note also that if $\lambda = 0$, which corresponds with no distortion, we recover the risk-neutral price.

Similar arguments lead to the following closed-form expressions for bid and ask pricing under the Wang transform:

$$bidEC^{WANG}(K, T; S_0, r, q, \sigma; \lambda) = EC(K, T; S_0, r, q + \lambda\sigma/\sqrt{T}, \sigma)$$
$$askEC^{WANG}(K, T; S_0, r, q, \sigma; \lambda) = EC(K, T; S_0, r, q - \lambda\sigma/\sqrt{T}, \sigma)$$
$$bidEP^{WANG}(K, T; S_0, r, q, \sigma; \lambda) = EP(K, T; S_0, r, q - \lambda\sigma/\sqrt{T}, \sigma)$$
$$askEP^{WANG}(K, T; S_0, r, q, \sigma; \lambda) = EP(K, T; S_0, r, q + \lambda\sigma/\sqrt{T}, \sigma).$$

5.4.3 Conic Black–Scholes Implied Volatility

Given some market bid and ask prices for a particular vanilla option, and given a distortion function with its parameter(s), one can calculate the volatility parameter needed in the above pricing formulas (in Equations (5.7)–(5.10)) to match the given market prices (under the Black–Scholes setting). We call these volatilities the *conic Black–Scholes implied volatilities*. Note that these are different from the traditional Black–Scholes implied volatilities calculated on the basis of the given bid or ask price. The latter do not depend on the distortion applied and, since the bid is below the ask price, we traditionally have a Black–Scholes implied volatility for the bid below the traditional Black–Scholes implied volatility for the ask. In the conic setting this is not necessarily the case any more, as is illustrated in the next Example.

Example 5.6 *Assume we have the following bid and ask quotes for an ATM call option on a stock with maturity $T = 1$:*

$$bid = 7.50 \text{ and } ask = 9.50,$$

where $S_0 = 100$, $r = 0.02$, $q = 0.01$. *Then the traditional Black–Scholes implied volatilities would equal* 17.83% *and* 22.95% *for the bid and the ask, respectively. The mid would have an implied vol of* 20.39%.

However, under a Wang transform with distortion parameter $\lambda = 0.1$, *we get*

$$bid = bidEC^{WANG}(K, T; S_0, r, q, \sigma_{bid}; \lambda) = 7.50$$
$$ask = askEC^{WANG}(K, T; S_0, r, q, \sigma_{ask}; \lambda) = 9.50$$

for $\sigma_{bid} = 20.62\%$ *and* $\sigma_{ask} = 19.98\%$. *These are the conic Black–Scholes implied volatilities for a Wang transform with distortion parameter* $\lambda = 0.1$.

5.5 Conic Vanilla Pricing

A similar analysis can be made under advanced models. Essential in the calculation of the vanilla's bid and ask price is an expression for the cdf of the stock price $F_S(x) = P(S_T \leq x)$. For the call's bid price we more precisely need:

$$\begin{aligned} F_{(S_T-K)^+}(x) &= 1 - P(S_T > K + x) \\ &= F_S(K + x), \qquad x \geq 0. \end{aligned}$$

For the call's ask price, we need

$$\begin{aligned} F_{-(S_T-K)^+}(x) &= P(S_T > K - x) \\ &= 1 - F_S(K - x), \qquad x \leq 0. \end{aligned}$$

Similarly, for the put's bid price we need access to

$$\begin{aligned} F_{(K-S_T)^+}(x) &= P(S_T \geq K - x) \\ &= 1 - F_S(K - x), \qquad 0 \leq x \leq K, \end{aligned}$$

and finally we need, for the put's ask price,

$$\begin{aligned} F_{-(K-S_T)^+}(x) &= 1 - P(S_T > K + x) \\ &= F_S(K + x), \qquad -K \leq x \leq 0. \end{aligned}$$

This leads to the following formulas that hold in general:

$$bidEC = \exp(-rT)\int_K^{+\infty} (x - K)\mathrm{d}\Psi(F_S(x))$$

$$askEC = \exp(-rT)\int_K^{+\infty} (K - x)\mathrm{d}\Psi(1 - F_S(x))$$

$$bidEP = \exp(-rT)\int_0^{K} (x - K)\mathrm{d}\Psi(1 - F_S(x))$$

$$askEP = \exp(-rT)\int_0^{K} (K - x)\mathrm{d}\Psi(F_S(x)).$$

Example 5.7 *We illustrate the conic call and put pricing in this section for the VG model and have*

$$\begin{aligned} F_S(x) = P(S_T \leq x) &= P\left(S_0 \exp((r-q+\omega)T + X_T) \leq x\right) \\ &= P\left(X_T \leq \log(x/S_0) - (r-q+\omega)T\right) \\ &= P(X_T \leq z_1(x)) \\ &= F_{VG}(z_1(x); \sigma\sqrt{T}, \nu/T, \theta T) \end{aligned}$$

where

$$z_1(x) = \log(x) - \log(S_0) - (r-q+\omega)T.$$

The cdf F_{VG} can be calculated, for example, via the VG-Laguerre approximation as in Section 3.1.3.

As an example, we reprice the European Put of Example 5.4. The put had one year to maturity and strike 90, while $S_0 = 100$, $r = 0.02$, $q = 0.01$. We now price it under the VG model with parameters $\sigma = 0.2$, $\nu = 0.75$ and $\theta = -0.3$. We use again the MINMAXVAR distortion with parameter $\lambda = 0.1$. In Figure 5.16, the cdf $F_{(K-S_T)^+}(x)$ and distorted cdf $\Psi_\lambda(F_{(K-S_T)^+}(x))$ of the put payoff is shown; the cdf is calculated using the VG-Laguerre approximation (3.4) using $n = 32$ node points. We observe that the probability of getting a zero payoff equals $F_{(K-S_T)^+}(0) = 0.6854$ under no distortion and $\Psi_\lambda(F_{(K-S_T)^+}(0)) = 0.7431$

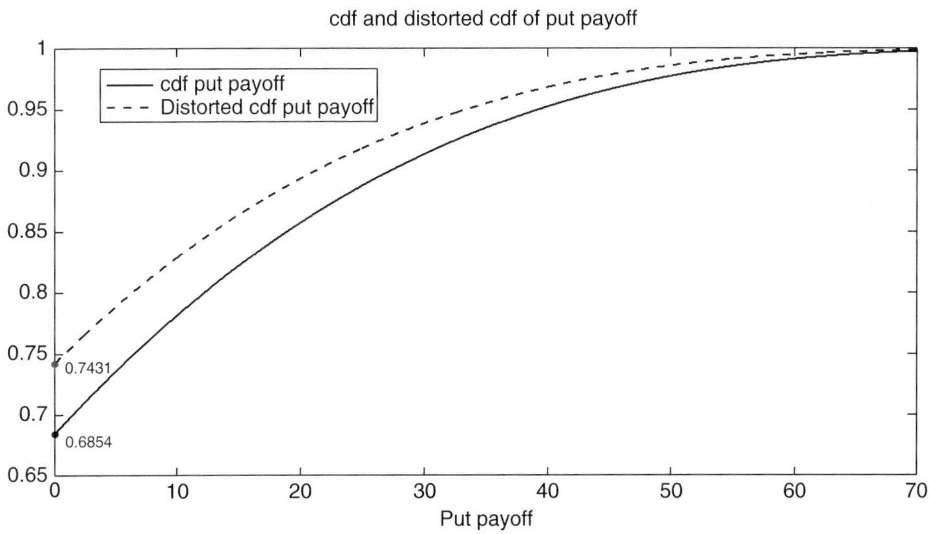

Figure 5.16 cdf and distorted cdf of a 90-strike put payoff under MINMAXVAR ($\lambda = 0.1$) under the VG setting with $S_0 = 100$, $r = 0.02$, $q = 0.01$, $T = 1$ and $\sigma = 0.20$, $\nu = 0.75$, $\theta = -0.3$

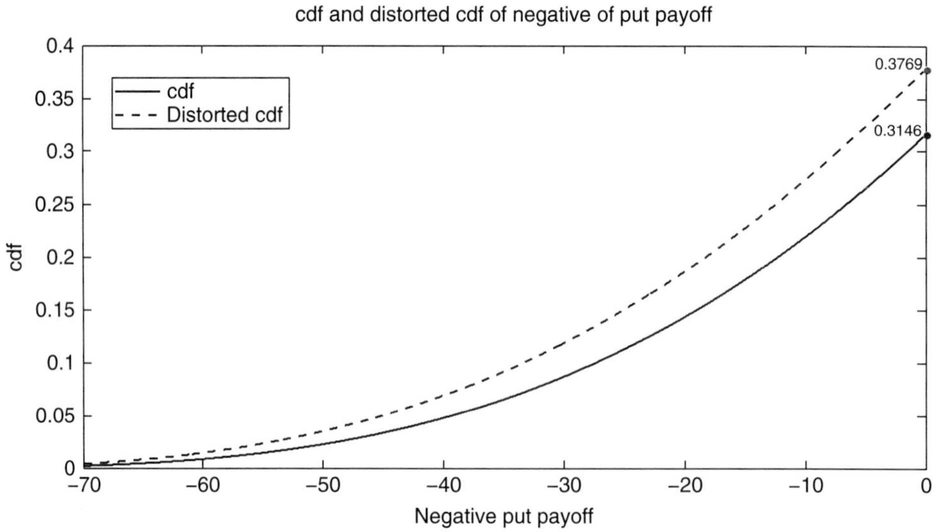

Figure 5.17 cdf and distorted cdf of the negative of a 90-strike put payoff under MINMAXVAR ($\lambda = 0.1$) under the VG setting with $S_0 = 100$, $r = 0.02$, $q = 0.01$, $T = 1$ and $\sigma = 0.20$, $\nu = 0.75$, $\theta = -0.3$

after distortion. Similarly, in Figure 5.17, we graph the cdf and the distorted cdf of the negative of the put payoff. On this figure one observes that the probability of getting a non-zero negative payoff equals $F_{-(K-S_T)^+}(0) = 0.3146$ *under no distortion and* $\Psi_\lambda(F_{-(K-S_T)^+}(0)) = 0.3769$ *after distortion.*

We now obtain 5.02 for the bid and 8.62 for the ask, while the risk-neutral price equals 6.65.

6

Applications of Conic Finance

In this chapter we provide several applications of the conic finance theory that we set up in the previous chapter. We first introduce the concept of implied liquidity inspired by the indexes of acceptability. Next, we illustrate how the counter-intuitive effects of booking profits due to your own credit deterioration (also referred to as Debt Valuation Adjustment or DVA) are mitigated under the conic bid and ask pricing theory. Finally, we illustrate how systematic risk in general or in a particular sector can be measured by exploring the sub-additivity property and, more precisely, by comparing bid and ask prices of vanillas in real markets with their values under a comonotonic setting.

6.1 Implied Liquidity

Assume the setting of Section 5.4, but let's look at the situation the other way around and fix a particular one-parameter distortion function family, like the MIN-MAXVAR distortion. Assume, given a bid and an ask price, $bidEC$ and $askEC$ respectively, an interest rate r and a dividend yield q. For an estimated volatility σ, for example, the implied volatility of the mid-price, one can calculate, using Formula 5.7, the λ-parameter that gives a perfect match with the in-the-market observed bid price. Let's denote this parameter with λ_{bid}. Similarly, one can calculate λ_{ask} using Formula 5.8. In Figure 6.1, the bid and ask prices are calculated as a function of λ.

Alternatively, one could say that the index of acceptability of the cash flow $(S_T - K)^+ - \exp(rT)bidEC$ equals λ_{bid} and similarly that the level of acceptability of the cash flow: $\exp(rT)askEC - (S_T - K)^+$ equals λ_{ask}.

Most of the time both λ_{bid} and λ_{ask} are very close to each other, and actually the parameter λ, which matches simultaneously as well as possible (in the least-squared sense), was called the implied liquidity parameter in Corcuera et al. (2012). It gives an indication of the market's liquidity: the higher the λ, the wider the spread and hence the less the liquidity. In Figure 6.2 the implied liquidity parameter is

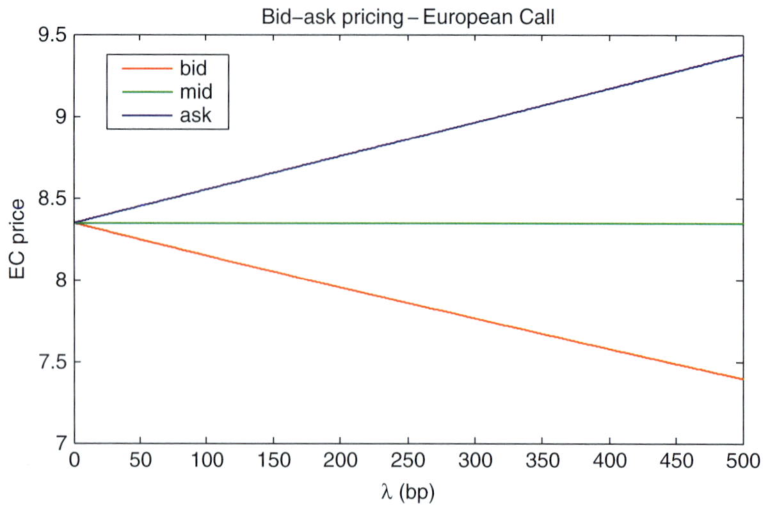

Figure 6.1 Bid and ask prices for a European Call as a function of the distortion parameter

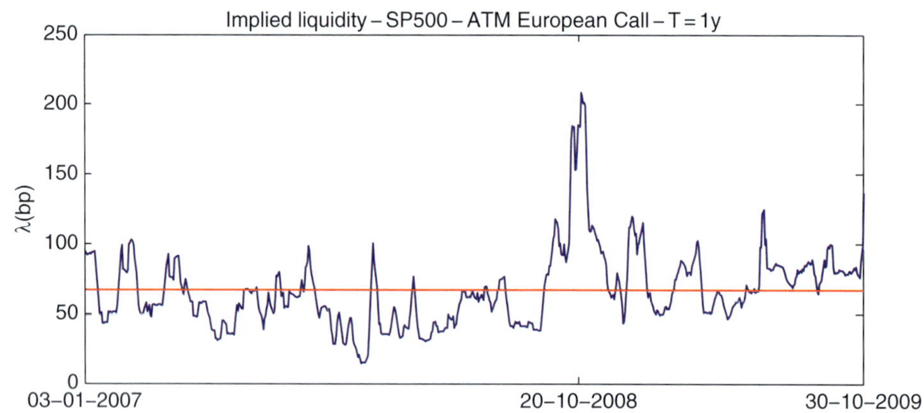

Figure 6.2 Implied liquidity

calculated on a daily basis for a European Call on the SP500 with strike as close as possible to the ATM level and with a maturity as close as possible to one year. The peak around the end of 2008 illustrates the dry-up in liquidity during the credit crunch in the last quarter of 2008. Further details can be found in Albrecher et al. (2013), Corcuera et al. (2012) and Guillaume (2015).

6.2 Conic CVA and DVA

Let us consider a very simple example (see also Madan and Schoutens (2016) and Madan (2015)). You have a unit exposure to a counterparty, which is due at time T. If your counterparty defaults, it returns a recovery value $0 \leq R \leq 1$. Assume that the risk-neutral default probability equals $0 < p < 1$ and that hence the risk-neutral no-default probability equals $1 - p$. Actually, you can view such an exposure as a defaultable zero-coupon bond (ZCB) maturing at time T. In the one-price framework, the value of this exposure/ZCB equals:

$$ZCB = \exp(-rT)\,(pR + (1 - p))\,,$$

since there are only two possible payouts (compare with a binomial tree setting): a payout of R with probability p and a payout of 1 with probability $1 - p$. If, for example, $T = 1$, $r = 0.01$, $R = 0.20$ and $p = 0.02$, then $ZCB = 0.9742$. Note that the price of a non-defaultable ZCB equals $\exp(-rT) = 0.99$.

In a two-price world, the default and no-default probabilities are distorted by a distortion function $\Psi(u)$. For the bid price, we first distort the probability with the worst payout and thus have a distorted default probability $\Psi(p)$. The distorted no-default probability then equals $1 - \Psi(p)$. Therefore

$$bidZCB = \exp(-rT)\,(\Psi(p)R + (1 - \Psi(p)))\,.$$

Employing the MINMAXVAR2 distortion function with a distortion parameters $\lambda = 0.25$ and $\gamma = 2$, $\Psi(p) = 0.1255$ and $bidZCB = 0.8906$.

In order to calculate the ask price, we focus on the negative of the payoff and then the worst payoff is the no-default payoff. If we distort the associated no-default probability we get $\Psi(1 - p) = 1$ and as a consequence have a distorted default probability equal to $1 - \Psi(1 - p) = 0$. The ask price of the defaultable ZCB is hence given by

$$askZCB = \exp(-rT)\,((1 - \Psi(1 - p))R + \Psi(1 - p))\,.$$

For the same parameter setting as above, we get $askZCB = 0.9900$.

CVA stands for the *Counterparty Value Adjustment* and reflects the adjustment in price due to default of the counterparty. Hence, assume you buy a ZCB from a counterparty and, as assumed above, the counterparty has a non-zero default probability p, then CVA equals the difference between the non-defaultable value and the defaultable value. DVA stands for *Debt Value Adjustment* and actually reflects an adjustment in price because the party selling the derivative potentially might not honour its obligations. In our example, the buyer of the zero coupon bond would make a counterparty value adjustment and book the bond, which is an asset for the buyer, at the default-free value minus the CVA. For the seller, the bond is booked as a liability, at the default free value minus the DVA.

In the one-price world, using the above parameter setting, the buyer would book the bond as an asset at 0.9742 and hence we have that the CVA equals 0.0158. The seller books the bond as a liability at 0.9742 and hence uses a DVA of 0.0158. In the one-price world CVA equals DVA.

The situation is different in a two-price setting. Prudent accounting would mean that assets are booked at the bid value, because it is this value the investor would get if he were to exit immediately. Similarly, liabilities would be booked at the ask value, because if one needs to close the position, one must buy back the asset, and this can be done at the ask value. Therefore, the buyer would book the bond as an asset at 0.8906 and hence we have that the CVA equals 0.0994. The seller books the bond as a liability at 0.99 and hence uses a zero DVA!

DVA has been criticized a lot because of the counter-intuitive effect that firms can book profits due to their own credit deterioration. Indeed, let's recalculate the above prices, but now for a default probability of double the original value, i.e. now $p = 0.04$. We get

$$ZCB = 0.9584, \quad bidZCB = 0.8225, \quad askZCB = 0.9900.$$

For the seller, the liability is in the one-price setting, now valued at 0.9584 instead of 0.9742. The seller is hence booking a DVA-profit of about 0.0158! In the two-price world, the profit for the seller due to its own credit deterioration is zero, since the ask has not moved and is still at 0.9900, leaving unchanged all other parameters.

6.3 Measuring Systematic Risk

Assume we are considering a weighted sum of n risks:

$$X = \sum_{i=1}^{n} w_i X_i.$$

One can think of this as a weighted basket of stocks composing an index or (sector) ETF, where X and X_i, $i = 1, \ldots, n$ are respectively denoting the index/ETF/basket value and the underlying stock price values at a given future date: T, say. We investigate the bid ask spread on this basket:

$$c(X) = \text{ask}(X) - \text{bid}(X) = \exp(-rT)\left(\sup_{Q\in\mathcal{M}} E_Q[X] - \inf_{Q\in\mathcal{M}} E_Q[X]\right)$$

for a non-empty set of probability measure $\mathcal{M}$. This spread is also referred to in Carr et al. (2011) as the capital requirement of the basket (see also Madan (2016b)). A property of such capital requirements is that the required capital of a portfolio equals the sum of the required capital over all the components when we have zero diversity or, in other words, full comonotonicity. More generally, the capital required for a portfolio is dominated by the sum of the individual requirements (due to the comonotone additivity in Section 4.4):

$$c(X) = c\left(\sum_{i=1}^{n} w_i X_i\right) \leq \sum_{i=1}^{n} w_i c(X_i),$$

with

$$c(X_i) = \text{ask}(X_i) - \text{bid}(X_i) = \exp(-rT)\left(\sup_{Q\in\mathcal{M}} E_Q[X_i] - \inf_{Q\in\mathcal{M}} E_Q[X]\right).$$

Under a perfect systematically driven market, or, in other words, under a perfect comonotonic setting (with zero diversity), one can show that both need to coincide: $c(X) = \sum_{i=1}^{n} w_i c(X_i)$. Under a non-perfect comonotone setting, however, the capital on the basket can be shown to be always less than the sum of the individual requirements: $c(X) < \sum_{i=1}^{n} w_i c(X_i)$. Hence, a measure of how comonotone markets currently are, or a measure of how (un)diversified the basket is, is the comparison of both; we compute the ratio of the basket capital requirement and the sum of the individual requirements:

$$0 \leq \text{DIVERSITY INDEX} = \frac{c\left(\sum_{i=1}^{n} w_i X_i\right)}{\sum_{i=1}^{n} w_i c(X_i)} \leq 1.$$

A value of 1 then represents a fully systematic/comonotone setting and no diversity.

As an example, one may take any sector ETF and its components to assess diversity in that sector, or one can take a stock index like the EURO STOXX 50, the Dow Jones Industrial Average, etc. We illustrate the above in the next example on the basis of the Dow Jones Industrial Average.

For an explicit computation we need a distribution for the risk, and one can obtain this by fitting a risk-neutral model to the option surface. Our systematic risk measure is then based on the derivatives market and as such is a forward-looking measure. It does not make use of historical time series data, but is based solely on the current market quotes. Further details can be found in Madan and Schoutens (2013).

Example 6.1 *A stock market index, such as the Dow Jones Industrial Average (DJI), for example, can be seen as a basket. Further, an option surface is typically available not only on the DJI itself, but also on each or at least most of its components. (If we do not have an option surface for a component, we work with an approximation of the index by excluding that name and reweighting the other names.) We hence assume that we can calibrate a given model on the DJI and on*

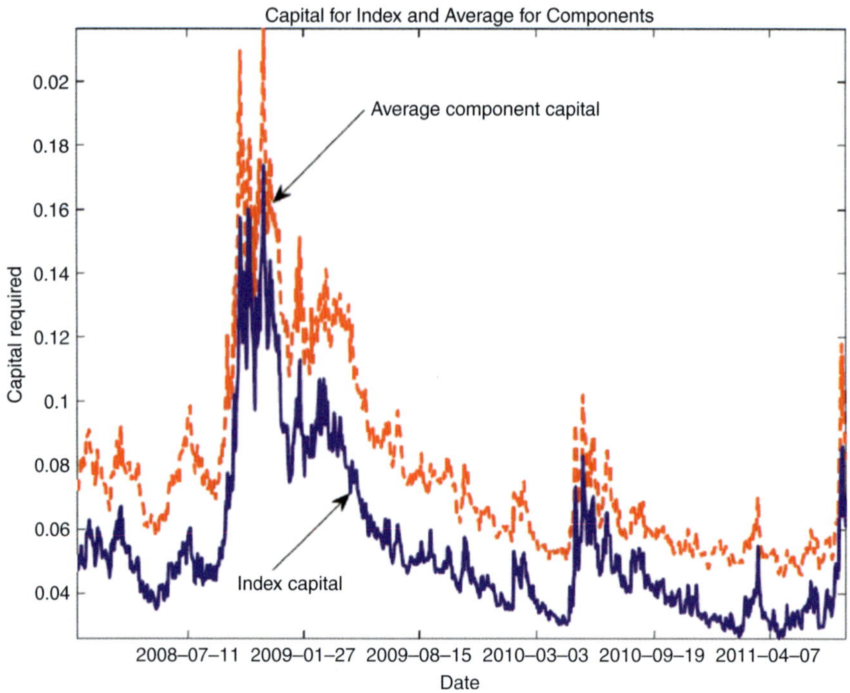

Figure 6.3 Dow Jones Industrial Average capital and average component capital

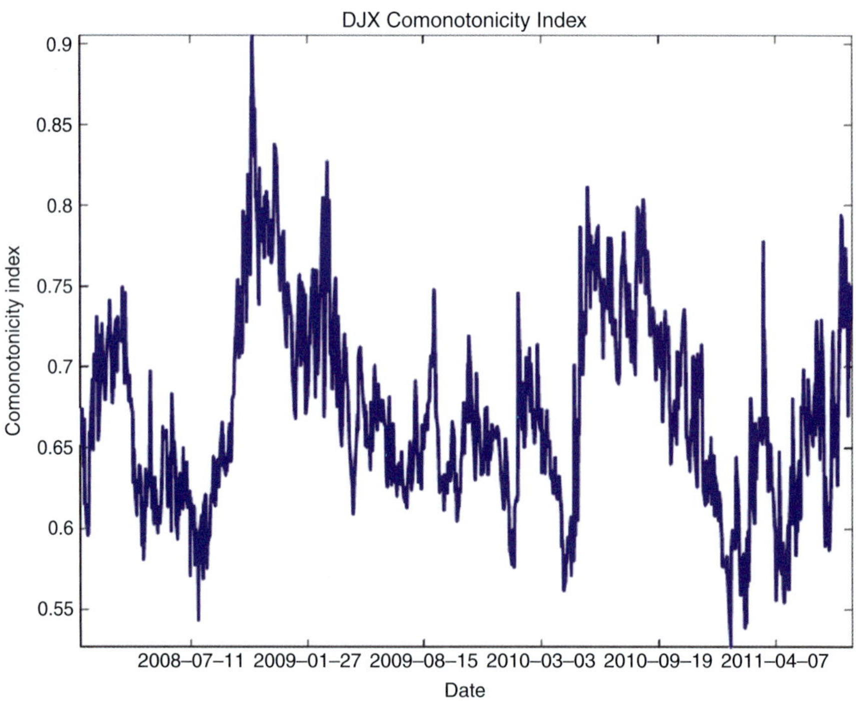

Figure 6.4 Diversity index as a measure of systematic risk

all the components to access the risk-neutral distribution. As illustration, we calibrate on a daily basis the so-called Sato model (see Section 2.6). The model has four parameters that fit the option surface adequately each day.

The random variable for which we charge capital are a unit forward contract on the index and the underlying stocks for a one-month maturity. We work here with the MINMAXVAR distortion and a $\lambda = 0.25$.

The diversity index is calculated daily for the time period 1 January 2008 to 16 August 2011. Figure 6.4 shows the capital required on holding the DJI for a month, along with the average capital required on the components. Figure 6.3 presents the ratio. This ratio is nothing other than the diversity index, and is unity when diversity is absent and markets are driven by a common systematic factor. For the DJI we observe that the diversity index peaked at 0.9 during the 2008 financial crisis.

7
Conic Portfolio Theory

In this chapter we use conic finance to calculate optimal portfolio compositions (see also Madan (2016a) and Eberlein and Madan (2015)). More precisely, we formulate a portfolio design problem with respect to maximizing a diversity measure. The diversity measure used is in terms of a conservative market value or bid price for the portfolio. Optimal portfolios maximizing diversification benefits are determined for long-only, long-short and volatility-constrained portfolios.

We also reformulate the problem in terms of an ask price and obtain a convex positive function bounded below by zero, which can be interpreted as an efficient (mean return ask price) frontier. The value-maximizing portfolio is then obtained at unit tangency with this frontier. Conic portfolio theory hence presents a value-based risk–reward framework in which both risk and reward are measured in the same units.

Consider n (dependent) assets $A^{(i)}$, $i = 1, \ldots, n$ and consider an investment horizon T. Let us denote the jointed centred, i.e. zero-mean, returns over this horizon, by $(R_i, i = 1, \ldots, n)$. We will consider a portfolio with weights a_i, $i = 1, \ldots, n$, representing the investment in the respective assets $A^{(i)}$. For the implementation part it will be important to be able to simulate the price behaviour of these assets efficiently.

7.1 The Optimal Conic Long-Only Portfolio

We first focus on the non-leveraged long-only portfolio, i.e. for which the positions are non-negative and sum to unity:

$$a_i \geq 0, \quad i = 1, \ldots, n \text{ and } \sum_{i=1}^{n} a_i = 1.$$

The centred portfolio return then equals

$$R_p = \sum_{i=1}^{n} a_i R_i.$$

Next, we observe that

$$\text{bid}\left(\sum_{i=1}^{n} a_i R_i\right) - \sum_{i=1}^{n} a_i \text{bid}(R_i) = \text{bid}(R_p) - \sum_{i=1}^{n} a_i \text{bid}(R_i)$$

is a diversity measure of the portfolio. Indeed, under a comonotonic setting the difference equals zero and in general $\text{bid}\left(R_p\right) \geq \sum_{i=1}^{n} a_i \text{bid}(R_i)$, and the more diversified the portfolio is, the bigger this gap. Further, we observe that, since the returns R_p and R_i have a zero mean (and hence have a zero risk-neutral price), $\text{bid}(R_p) \leq 0$ and $\text{bid}(R_i) \leq 0$, $i = 1, \ldots, n$.

Since in general $\text{ask}(X) = -\text{bid}(-X)$, the function $\tilde{c}(a) = -\text{bid}\left(R_p\right)$ can actually also been seen as the positive ask price for the negative of the portfolio return: $\tilde{c}(a) = \text{ask}\left(-R_p\right)$. The function $\tilde{c}(a)$ is a convex positive function bounded below by zero. Hence, if we write $\mu_p = -\sum_{i=1}^{n} a_i \text{bid}(R_i) = \sum_{i=1}^{n} a_i \text{ask}(-R_i)$, the diversity gap is decomposed into the reward μ_p and the risk $\tilde{c}(a)$ measured by the ask price. We have

$$\mu_p \geq \tilde{c}(a) \geq 0.$$

We can write the diversity gap as

$$\text{bid}\left(R_p\right) - \sum_{i=1}^{n} a_i \text{bid}(R_i) = -\text{ask}\left(-R_p\right) + \sum_{i=1}^{n} a_i \text{ask}(-R_i) = \mu_p - \tilde{c}(a).$$

Hence, in general the diversity gap may be seen as a risk–reward trade-off where reward μ_p is given by the negative of the mean of the bid of the returns and the risk $\tilde{c}(a)$ is modelled by the ask price for the negative return. The risk and reward here are in the same units and trade one to one. In classical mean variance approaches to risk and reward, we have the unnatural situation of measuring the two in different units such as dollars and squared dollars (or drifts and variances).

Another attractive feature of this formulation of the problem is that, by construction, $\tilde{c}(a)$ is homogeneous of order one in a:

$$\tilde{c}(\alpha a) = \alpha \tilde{c}(a)$$

or that the total risk is made up of the sum of marginal risk contributions times dollar exposures.

In that perspective, we can now maximize over all positions a_i, $i = 1, \ldots, n$ this gap. This can be achieved, for example, via a Monte Carlo simulation of the jointed centred returns. More precisely, one may generate, for any candidate portfolio with allocations a_i, $i = 1, \ldots, n$, a sample of portfolio returns $R_{p,m}$, $m = 1, \ldots, M$ for M simulations of the individual centred asset returns. The bid prices $\text{bid}(R_p)$

and $\text{bid}(R_i)$ may then be evaluated from the ordered returns $R_{p,(m)}$ and $R_{i,(m)}$, $m = 1, \ldots, M$, arranged in increasing order, by

$$\text{bid}(R_p) = \sum_{m=1}^{M} R_{p,(m)} \left(\Psi\left(\frac{m}{M}\right) - \Psi\left(\frac{m-1}{M}\right) \right) \tag{7.1}$$

and

$$\text{bid}(R_i) = \sum_{m=1}^{M} R_{i,(m)} \left(\Psi\left(\frac{m}{M}\right) - \Psi\left(\frac{m-1}{M}\right) \right) \tag{7.2}$$

for a distortion Ψ (compare with Equation (5.3)).

For a practical implementation the above relies on the ability to simulate centred returns. Here, different approaches can be followed. The most simple one is a bootstrapping technique, where, out of a historical sample, new returns are sampled. Alternatively, one can employ multivariate models estimated on either historical joint data series or on risk-neutral derivatives data.

We illustrate first the bootstrap technique on a portfolio of $n = 5$ stocks, with tickers: GE, INTC, IBM, JNJ, AAPL. We use as input daily price observations from 3 January 2000 to 17 August 2009 and use a one-month time horizon. To calculate monthly return samples from daily return samples, we will essentially multiply k daily returns, where $k = 20$, which is the number of trading days in the given horizon. This can be criticized and more realistic schemes applied, but for the sake of the illustration it will do.

For a given candidate portfolio with allocations a_i, $i = 1, \ldots, n$, we more precisely perform the following steps:

1. We calculate for each stock on each day the daily return in our time series (except for the first day).
2. We now bootstrap from that time series: we perform M times the following operations. We select randomly $k = 20$ days and consider the corresponding daily returns $(r_{i,j}, i = 1, \ldots, n)$, $j = 1 \ldots, k$. For the mth run, $m = 1, \ldots, M$, the asset returns over our investment horizon of one month are calculated as $\tilde{R}_{i,m} = \prod_{j=1}^{k}(1 + r_{i,j}) - 1$.
3. We then centre these M returns. The centred monthly return for the ith stock is calculated as

$$R_{i,m} = \tilde{R}_{i,m} - \frac{1}{M} \sum_{m=1}^{M} \tilde{R}_{i,m}, \qquad i = 1, \ldots, n, m = 1, \ldots, M.$$

4. We calculate for the candidate portfolio with allocations a_i, $i = 1, \ldots, n$, the centred portfolio returns

Table 7.1 *Conic optimal long-only portfolios.*

Stress	GE	INTC	IBM	JNJ	AAPL
0.1000	0.1807	0.1765	0.0967	0.3157	0.2304
0.2000	0.1778	0.1772	0.0965	0.3165	0.2320
0.3000	0.1751	0.1778	0.0962	0.3170	0.2339
0.4000	0.1726	0.1782	0.0958	0.3173	0.2360
0.5000	0.1703	0.1786	0.0955	0.3174	0.2382
0.6000	0.1681	0.1790	0.0953	0.3172	0.2405
0.7000	0.1660	0.1792	0.0954	0.3166	0.2428
0.8000	0.1640	0.1794	0.0958	0.3158	0.2450
0.9000	0.1620	0.1794	0.0967	0.3146	0.2473
1.0000	0.1602	0.1794	0.0979	0.3129	0.2496

Table 7.2 *The diversity gaps, portfolio means and the negative of the portfolio's bid prices.*

Stress	$\text{bid}\left(R_p\right) - \sum_{i=1}^{n} a_i \text{bid}(R_i)$	μ_p	$\tilde{c}(a)$
0.1000	0.0053	0.0182	0.0129
0.2000	0.0102	0.0352	0.0250
0.3000	0.0149	0.0512	0.0363
0.4000	0.0194	0.0665	0.0471
0.5000	0.0239	0.0812	0.0573
0.6000	0.0283	0.0955	0.0672
0.7000	0.0327	0.1094	0.0767
0.8000	0.0370	0.1230	0.0860
0.9000	0.0414	0.1363	0.0950
1.0000	0.0457	0.1494	0.1037

$$R_{p,m} = \sum_{i=1}^{n} a_i R_{i,m}, \qquad m = 1, \ldots, M.$$

5. Using Equations (7.1) and (7.2), we calculate the diversity measure of the candidate portfolio as

$$\mu_p - \tilde{c}(a) = \text{bid}(R_p) - \sum_{i=1}^{n} a_i \text{bid}(R_i).$$

We optimize over the allocations a_i, $i = 1, \ldots, n$, by performing a constrained optimization procedure. The applicable constraints are $a_i \geq 0$, $i = 1, \ldots, n$ and $\sum_{i=1}^{n} a_i = 1$.

For the MINMAXVAR distortion with stress levels from 0.1 to 1.0, we obtain the following optimal long-only conic portfolio as in Table 7.1. The corresponding diversity gaps $\text{bid}(R_p) - \sum_{i=1}^{n} a_i \text{bid}(R_i)$, portfolio means $\mu_p = -\sum_{i=1}^{n} a_i \text{bid}(R_i)$ and the negative of the portfolio's bid prices $\tilde{c}(a)$ are given in Table 7.2.

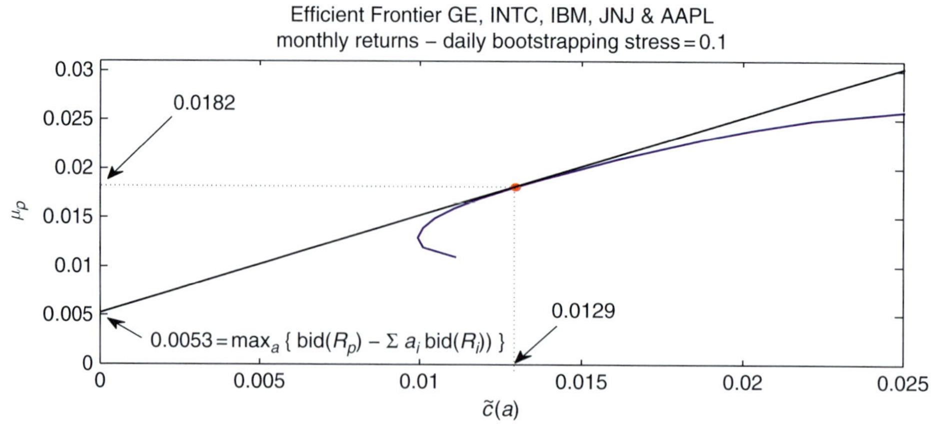

Figure 7.1 Efficient frontier

7.2 The Conic Efficient Frontier

As both the reward and risk scale, there is an efficiency frontier defined by the minimum $\tilde{c}(a)$ subject to $a_i \geq 0$, $\sum a_i = 1$ and a given μ_p. This is a well-defined problem as $\tilde{c}(a)$ is bounded below by zero and it may be solved for an interior solution.

This may graphically be represented in the traditional way, but now with the value of $\tilde{c}(a) = -\text{bid}\left(R_p\right)$ on the x-axis and on the y-axis the value of $\mu_p = -\sum_{i=1}^{n} a_i \text{bid}(R_i)$. See Figure 7.1. This graph is calculated by performing a constrained optimization. We minimize $\tilde{c}(a)$ with the constraints

$$a_i \geq 0, \quad \sum_{i=1}^{n} a_i = 1 \text{ and } \mu_p = -\sum_{i=1}^{n} a_i \text{bid}(R_i,)$$

for a range of given μ_p values.

The optimal long-only portfolio is represented by the point where the tangent line with slope unity touches the efficient frontier. The intersect of this line with the y-axis is the diversity gap $\text{bid}\left(R_p\right) - \sum_{i=1}^{n} a_i \text{bid}(R_i)$.

7.3 The Conic Long-Short Portfolio with a Volatility Constraint

A similar exercise can be made for a long-short portfolio. For this unconstrained problem we may always construct the efficiency frontier as above, since the ask price minimization problem is well defined for a portfolio with a given mean. If the efficient frontier has a slope above unity for all μ_p then the unconstrained problem may not have a finite solution.

Suppose the unconstrained long-short problem is well posed. This may be checked by ensuring that mean returns are below those required for acceptability of the individual assets at the stress levels being employed. In this case one does not need to construct the mean ask price efficiency frontier and then locate the value maximizing portfolio on this frontier, as done above. One may directly numerically solve the convex optimization problem of maximizing the bid price for the portfolio subject to the constraint $\sum_{i=1}^{n} a_i = 1$.

Additionally, if the unconstrained problem is not well posed, one may solve for a volatility constrained investment by maximizing the bid price subject to the two constraints $\sum_{i=1}^{n} a_i = 1$ and $\sqrt{a'\Sigma a} \leq \sigma^*$. The covariance matrix Σ can easily be calculated from our sample and $\sigma^* > 0$ is our target volatility.

The allocations can now be positive or negative (and are no longer bounded). In addition, we impose that the resulting portfolio has to have a (monthly) volatility below σ^*. We illustrate this setting with another method of sampling. More precisely, we will employ a parametric model, namely the VG, in combination with moment matching parameter estimation of Section 3.2.2 and an Independent Component Analysis (ICA). ICA estimates the independent components from given multidimensional signals together with its mixing matrix.

More specifically, we consider a concrete portfolio of $n = 5$ stocks with tickers: PEP, MSFT, JNJ, XOM, ORCL. We use as input daily price observations from 3 January 2000 to 17 August 2009 and use a one-month time horizon. We run the procedure, using the MINMAXVAR distortion with stress levels 0.1 and use it in the long/short setting with a volatility constraint of 15% (yearly volatility): $\sigma^* = 0.15/\sqrt{12}$.

We perform the following steps and get the following intermediate results:

1. We calculate the daily log-returns for our underlying assets and centre them (i.e. subtract the individual means). Denote the corresponding matrix L. It has n columns and a number of rows corresponding to the number of days in our observation period. In our case there are 2419 trading days from 3 January 2000 to 17 August 2009 and hence we have $d = 2418$ log-returns.
2. We perform an ICA analysis and obtain the estimated independent components $\tilde{L}$ and the corresponding mixing matrix A. We have the following relation:

$$L = \tilde{L} * A,$$

where A is an n by n matrix and $\tilde{L}$ has n columns and d rows. For our data set we obtain,

$$A = \begin{bmatrix} -0.6700 & -0.7285 & -0.8098 & -0.9696 & -0.6817 \\ -0.0425 & 0.7786 & -0.0619 & 0.0015 & -0.0552 \\ 0.7591 & -0.0065 & -0.2520 & -0.1213 & -0.0870 \\ -0.0911 & -0.0160 & -0.3555 & 0.5913 & -0.1735 \\ 0.0085 & -0.3021 & 0.0601 & -0.0198 & -0.9404 \end{bmatrix}$$

Table 7.3 *Variance, skewness and kurtosis.*

Component	1	2	3	4	5
variance	0.00021999	0.00052874	0.00020380	0.00032343	0.0009986
skewness	0.00278738	−1.02190702	−0.08738758	0.00058489	−0.0671575
kurtosis	17.23245657	18.61010226	12.98913944	7.62267465	9.1911930

3. We perform a moment-matching estimation on each of our n independent components using the d observations present in each of the columns of the matrix $\tilde{L}$ and fit a VG distribution. The variance, skewness and kurtosis are given in Table 7.3. Employing the equation of Section 3.2.2 (note that we use $T = 1$ day), we obtain the VG parameters for each of our five components. Note that these parameters are expressed in a daily base ($T = 1$ day):

i	$\sigma_{1d,i}$	$\nu_{1d,i}$	$\theta_{1d,i}$
1	0.01483235	4.74415048	0.00000290
2	0.02272021	4.97038478	−0.00158844
3	0.01427424	3.32801605	−0.00012496
4	0.01798435	1.85408915	0.00000227
5	0.03159809	2.06272873	−0.00034298

4. Next, we have to simulate and model the monthly returns. Assuming independent and stationary increments, hence just simple VG dynamics (which again can be criticized, and maybe some more sophisticated dynamics are more realistic), we obtain the distribution of monthly returns. Indeed, using

i	$\sigma_{1m,i}$	$\nu_{1m,i}$	$\theta_{1m,i}$
1	0.06633228	0.23720752	0.00005809
2	0.10160788	0.24851923	−0.03176890
3	0.06383634	0.16640080	−0.00249929
4	0.08042846	0.07704457	0.00004551
5	0.14131097	0.10313643	−0.00685977

the properties of the VG distributions, we have that for each $i = 1, \ldots, n$ the parameter for monthly returns are given by

$$\begin{aligned} \sigma_{1m,i} &= \sigma_{1d,i}\sqrt{k} \\ \nu_{1m,i} &= \nu_{1d,i}/k \\ \theta_{1m,i} &= \theta_{1d,i}k, \end{aligned}$$

Table 7.4 *Conic optimal portfolio without and with vol constraint.*

	$\sqrt{12}\sqrt{a'\Sigma a}$		$\mu_p - \tilde{c}(a)$	μ_p	$\tilde{c}(a)$
Long only no vol constraint	19.46%		0.0084	0.0179	0.0095
Long/short with $\sigma^* = 0.15/\sqrt{12}$ constraint	15.00%		0.0083	0.0156	0.0073
	PEP	MSFT	JNJ	XOM	ORCL
Long only no vol constraint	0.3218	0.5118	0.0565	0.0937	0.0161
Long/short with $\sigma^* = 0.15/\sqrt{12}$ constraint	0.2645	0.4098	0.1687	0.2056	−0.0486

where $k = 20$ is used for the number of trading days in one month.

5. Using traditional random number simulation of VG variates (see Section 3.3.3), we now perform M times the following operations. For the mth run, $m = 1, \ldots, M$, denote the sampled one-month log-return for the ith asset by $r_{i,m}$, $i = 1, \ldots, n$. These are each sampled as $r_{i,m} = \theta_{1m,i}(g - 1) + \sigma_{1m,i}\sqrt{g}\epsilon$, where g is a Gamma$(1/\nu_{1m,i}, 1/\nu_{1m,i})$ random number and ϵ a standard normal variate. Note that due to the -1 in the first term the sampled result has a zero mean. Then the (relative) centred returns over our investment horizon of one month are

$$R_{i,m} = \frac{\exp(r_{i,m})}{\frac{1}{M}\sum_{m=1}^{M}\exp(r_{i,m})} - 1, \quad i = 1, \ldots, n.$$

As above, we next calculate for the candidate portfolio with allocations a_i, $i = 1, \ldots, n$, the centred portfolio returns

$$R_{p,m} = \sum_{i=1}^{n} a_i R_{i,m}, \qquad m = 1, \ldots, M.$$

6. Using Equations (7.1) and (7.2), we calculate the diversity measure of the candidate portfolio as

$$\mu_p - \tilde{c}(a) = \text{bid}(R_p) - \sum_{i=1}^{n} a_i \text{bid}(R_i).$$

We optimize over the allocations a_i, $i = 1, \ldots, n$, by performing a constrained optimization procedure. The applicable constraints are $\sum_{i=1}^{n} a_i = 1$ and $\sqrt{a'\Sigma a} \leq \sigma^*$.

We report the results for the long only procedure without a volatility constraint and the long/short setting with a volatility constraint of 15% (yearly volatility) and also provide the resulting portfolio's volatility (in yearly terms) and the diversity gap $\mathrm{bid}\left(R_p\right) - \sum_{i=1}^{n} a_i \mathrm{bid}(R_i)$, the portfolio's mean μ_p and the ask price $\tilde{c}(a)$ defined as above. Table 7.4 summarizes the results.

8
Conic Hedging

In this chapter we elaborate on a hedging methodology that we call *conic hedging* (see also Madan et al. (2016b)). Conic hedging is permitting educated hedging choices with wide applications. We employ bid price and ask price functionals to construct hedging strategies. The primary focus of these strategies is to adopt positions maximizing bid prices or minimizing ask prices for the one-step-ahead risk held or promised. Risk management and hedging then have a new market value-enhancing perspective which is different from the classical forms of risk mitigation, local variance minimization or even expected utility maximization.

We will seek to design hedging strategies to maximize the concave bid price for positions held or minimize the convex ask price for positions promised. For any set of chosen hedging instruments, such optimization problems are control problems related to non-linear valuation functionals.

Conic hedging procedures may be employed to hedge risks correlated with market indices that have liquid short maturity or weekly options trading when the underlying risk itself does not have such possibilities. This is further illustrated in Chapter 9, in the context of the hedging of equity-linked insurance contracts.

After introducing the concepts in a one-step tree setting, we move to a multi-step setting and make the procedure dynamic.

The concept of dynamic conic delta and gamma hedging introduced here differs fundamentally from risk management approaches aimed at delta and/or gamma neutrality. The latter seek to zero out certain derivatives of the current value function. We seek to add positions to future value functions that enhance current market values.

8.1 Conic Delta Hedging

The key idea of conic hedging is that we wrap the derivative in a package, which has the same risk-neutral price but a more competitive bid or ask price.

To be more precise, instead of pricing just the derivative, we will now calculate the bid and ask prices of a portfolio consisting of the derivative in combination with a stock position. We assume, for simplicity, there are no dividends. More concretely, we consider the portfolio of one derivative and a position of Δ forward contracts paying out at maturity T the amount $(S_T - \exp(rT)S_0)$. Note that the risk-neutral upfront price to be paid for such a forward contract is zero, and hence the risk-neutral price of the portfolio is unchanged.

The key idea of conic hedging is that we look for an optimal value, Δ_{conic}. Optimal here means having a maximal bid price or minimal ask price.

Let us first turn to the concrete settings of the binomial and trinomial tree as in Sections 5.1.1 and 5.1.3. Next, we will compare the traditional delta of the Black–Scholes model with its conic counterpart.

8.1.1 Conic Delta Hedging in a Binomial Tree

Under the binomial tree, the payoff in the up-state equals f_u, and it equals f_d in the down-state. Combine this derivative now with Δ forward contracts. Then this portfolio pays out $\pi_u = f_u + \Delta(u - \exp(rT))S_0$ in the up-state and $\pi_d = f_d + \Delta(d - \exp(rT))S_0$ in the down-state.

Assume for a given Δ that $\pi_u \geq \pi_d$; then we have that the bid price of this portfolio equals

$$\text{bid} = \exp(-r\Delta t)\left((1 - \Psi(1-p))\pi_u + \Psi(1-p)\pi_d\right),$$

where we recall that p is the risk-neutral probability of moving to the up-state. If we were to have for a given Δ that $\pi_u \leq \pi_d$, then we would have that the bid price of this portfolio equals

$$\text{bid} = \exp(-r\Delta t)\left(\Psi(p))\pi_u + (1 - \Psi(p))\pi_d\right).$$

Conic delta hedging looks for the Δ^{bid}_{conic}, such that the bid price is maximal. In Figure 8.1, we graph the bid price of the ATM call of Example 5.1 for different values of Δ.

We see that the maximal value is attained for $\Delta^{bid}_{conic} = -0.5144$; the corresponding bid price for this Δ^{bid}_{conic} equals 2.9264, which is actually the risk-neutral price. This is actually no surprise, since the traditional delta for a binomial tree equals $\Delta_{tree} = (f_u - f_d)/(S_0(u-d)) = 0.5144$, the negative of the conic delta calculated under this setting. Note that we have the negative since we have to *sell* Δ_{tree} stock to implement the traditional hedge.

For this delta-hedged position, the final portfolio payoff is risk-free and hence has the same value in both the up-state and down-state. Our cdf hence is degenerate with just a jump of size 1 in this payoff value. Distorting this cdf gives rise to the same cdf, since there is nothing to distort. The bid price hence equals the

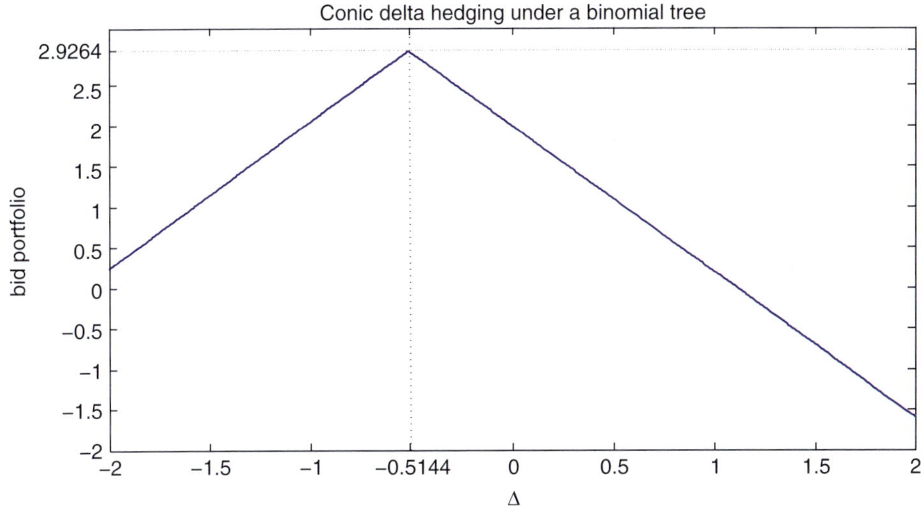

Figure 8.1 Conic delta hedging under a binomial tree

risk-neutral price. The same reasoning can be made for the ask price, and in this particular setting of the binomial tree, we get

$$\Delta^{bid}_{conic} = \Delta^{ask}_{conic} = -\Delta_{tree}.$$

8.1.2 Conic Delta Hedging in a Trinomial Tree

The situation is different, though, in the trinomial tree setting, which is an incomplete market setting. Now consider a derivative paying out

$$\pi_u = f_u + \Delta(u - \exp(rT))S_0 \text{ in the up-state;}$$
$$\pi_m = f_m + \Delta(m - \exp(rT))S_0 \text{ in the middle-state;}$$
$$\pi_d = f_d + \Delta(d - \exp(rT))S_0 \text{ in the down-state.}$$

Using the techniques of Section 5.1.3, we can calculate the bid price of this portfolio and look for the Δ^{bid}_{conic}, such that this bid price is maximized. In Figure 8.2, we graph the bid price of the setting of Example 5.2 for different values of Δ.

We see that the maximal value is attained for $\Delta^{bid}_{conic} = 0.0500$, the corresponding bid price equals for this Δ^{bid}_{conic} to 1.7808, which is below the risk-neutral price of 2.1649.

A similar exercise can be made for the ask price and one can look for the Δ^{ask}_{conic} value for which the portfolio has the lowest ask price. We see, as in Figure 8.3, that the minimal ask price is attained for $\Delta^{ask}_{conic} = 0.0500$; the corresponding ask price for this Δ^{ask}_{conic} equals 2.4922, which is above the risk-neutral price of 2.1649.

Table 8.1 *Conic hedging under trinomial tree for* $\Delta^{bid}_{conic} = \Delta^{ask}_{conic} = 0.0500$.

Unhedged bid	Conic delta hedged bid	Risk-neutral	Conic delta hedged ask	Unhedged ask
1.7326	1.7808	2.1649	2.4922	2.5136

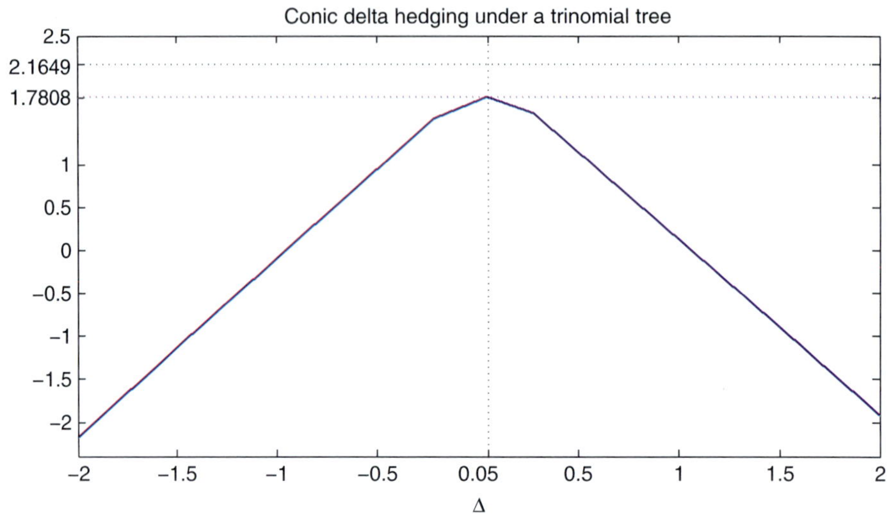

Figure 8.2 Conic delta hedging under a trinomial tree

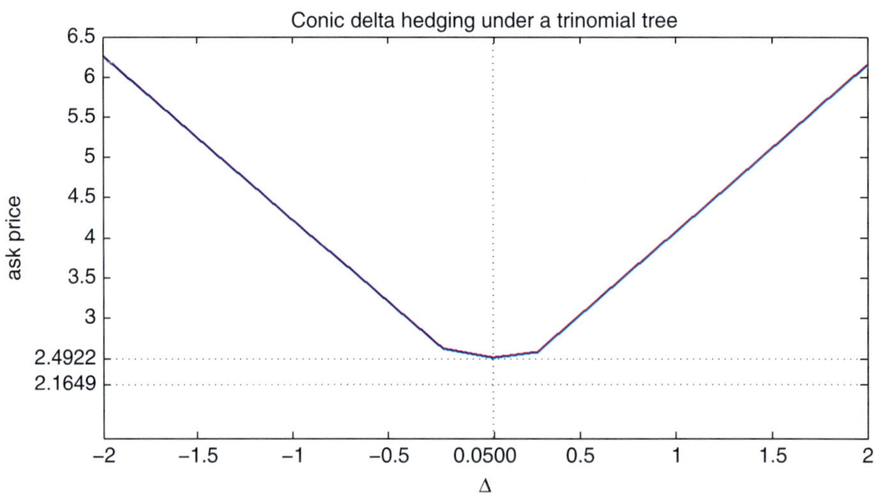

Figure 8.3 Conic delta hedging under a trinomial tree

We observe that the Δ^{bid}_{conic} giving rise to the maximal bid price equals Δ^{ask}_{conic} giving rise to the minimal ask price. This is a special feature of the trinomial model (and does not hold in general). In our situation, and for this particular value, the portfolio's payoff in the up-state and down-state are exactly the same, and conic delta hedging under a trinomial model brings us into a binomial setting.

8.1.3 Conic Delta Hedging under Black–Scholes

We can apply the same concept to any model and any derivative. We illustrate this under the Black–Scholes setting. In Figure 8.4, we have calculated the bid price for an ATM call option with maturity T, take equal to 1 month in the illustration, in combination with a number of forward contracts. More concretely, we consider the portfolio of one derivative and a position of Δ forward contracts paying out at maturity T the amount $(S_T - \exp(rT)S_0)$, with T the same maturity as our call option. We have plotted the package bid price for our different conic delta values Δ. We observe that the bid price reaches its maximal value for $\Delta^{bid}_{conic} = -0.527$. In Figure 8.5, we repeat the exercise but now minimizing the ask price. We observe that the ask price reaches its minimal value for $\Delta^{ask}_{conic} = -0.5330$.

In Figure 8.6, we compare for a wide range of strikes, the optimal Δ^{bid}_{conic}, our conic delta for the bid with the negative of the traditional delta ($\Delta_{BS} = N(d_1)$). Note again that we take the negative because, in order to traditionally delta-hedge a call option, one has to *sell* delta (Δ_{BS}) amount of stock.

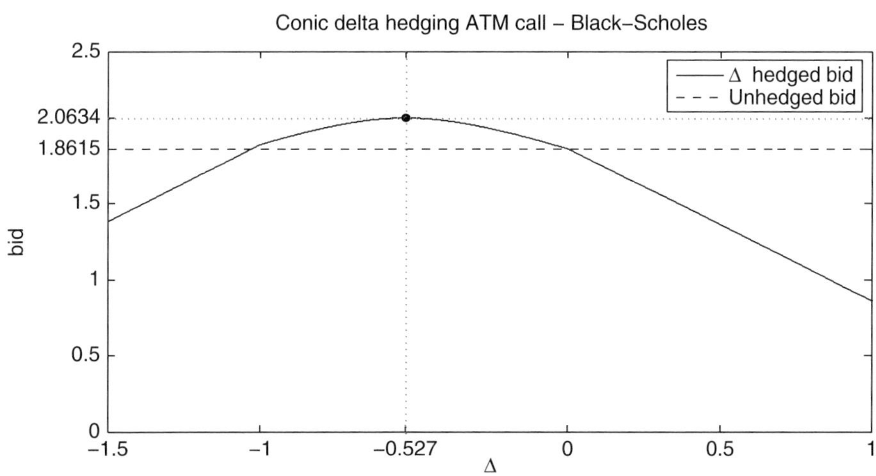

Figure 8.4 Conic delta for the bid price of an ATM call under the Black–Scholes model ($r = 0.01, q = 0, T = 1/12, S_0 = 100, \sigma = 0.2, \lambda = 0.10$)

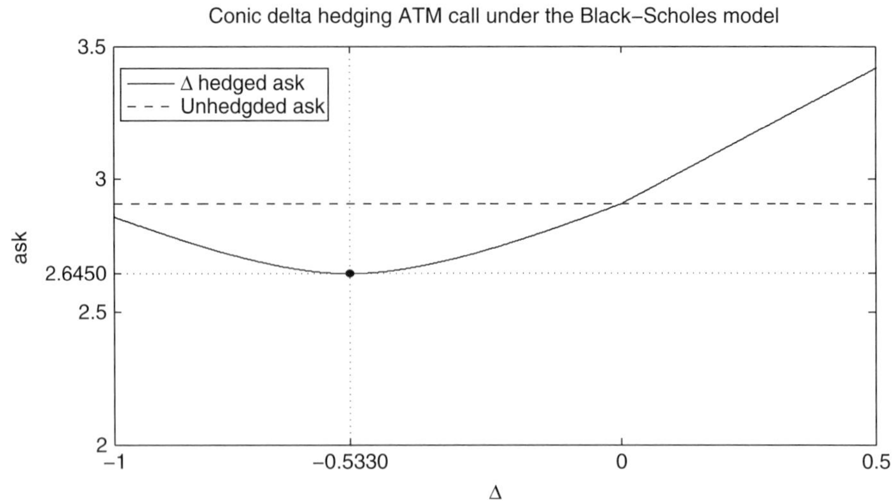

Figure 8.5 Conic delta for the ask price of an ATM call under the Black–Scholes model ($r = 0.01, q = 0, T = 1/12, S_0 = 100, \sigma = 0.2, \lambda = 0.10$)

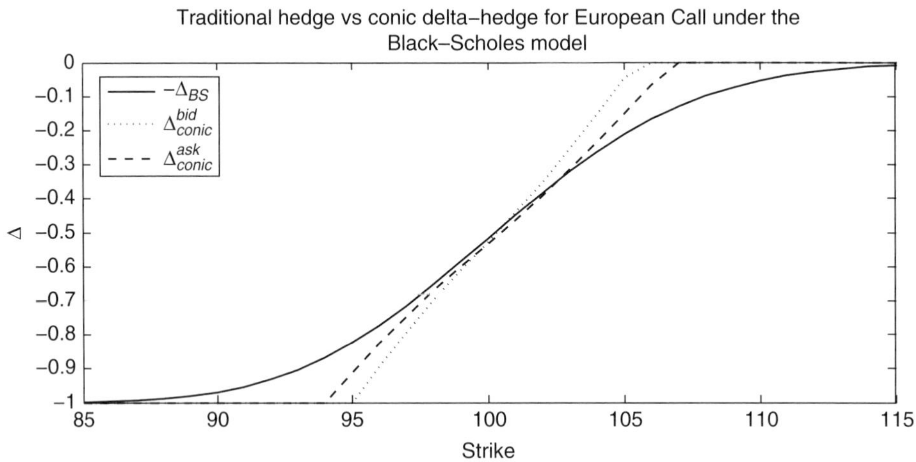

Figure 8.6 Traditional hedge vs conic delta-hedge for European Call under the Black–Scholes model ($r = 0.01, q = 0, T = 1/12, S_0 = 100, \sigma = 0.2, \lambda = 0.10$)

Although the traditional delta and the conic delta are completely different concepts, with the former operating very locally and on a small timescale or even instantaneously and the latter operating globally on the final maturity, their values are in quite comparable ranges. The conic hedging is essentially taking a market value-enhancing perspective, raising the bid price or lowering the ask price, while the risk-neutral price remains the same, which is different from the classical forms of risk mitigation. The traditional delta hedging seeks to zero out the first derivative of the current value function. Conic hedging seeks to add positions to future value functions that enhance current market values.

Finally, note that the conic delta is depending on the chosen distortion and distortion parameter.

8.2 Conic Delta-Gamma Hedging

One can expand the conic delta hedge by allowing for the possibility of investing in other instruments. One could (in addition), for example, allow for positions in a contract that pays at time T a quadratic payoff. We call such a hedging *conic delta-gamma hedging*. As an example, let us consider a swap contract that pays out at T the square of the realized price change less the forward price for this claim:

$$(S_T - S_0 \exp(rT))^2 - E_Q[(S_T - S_0 \exp(rT))^2].$$

Again, the upfront risk-neutral price of this quadratic claim is zero and hence a portfolio with our original derivative, combined with Δ forward contracts paying out $(S_T - \exp(rT)S_0)$ and Γ of the above quadratic contracts, has the same risk-neutral price as the derivative itself. However, the bid and the ask prices of this package are not the same as the bid and ask prices of the original derivative.

Using the techniques of Section 5.1.3, we can calculate the bid and ask price of such a portfolio and we can now perform a two-dimensional numerical optimization, looking for the optimal pairs $(\Delta^{bid}_{conic}, \Gamma^{bid}_{conic})$ and $(\Delta^{ask}_{conic}, \Gamma^{ask}_{conic})$ respectively maximizing the portfolio's bid price and minimizing the portfolio's ask price.

For our trinomial tree setting of Section 8.1.2 we observe that the maximal value is attained for $\Delta^{bid}_{conic} = 0.0334$ and $\Gamma^{bid}_{conic} = 0.0248$; see Figure 8.7. The corresponding portfolio's bid price equals 2.1649, which is the risk-neutral price. Actually, the ask price is also minimized for these values: $\Delta^{ask}_{conic} = 0.0334$ ans $\Gamma^{ask}_{conic} = 0.0248$. Due to similar reasoning as for the conic delta hedge in the binomial model, we have that under the trinomial model, conic delta-gamma eliminates all uncertainty in the payoff and therefore the portfolio's bid price equals the risk-neutral price. This is not true under all models, however.

8.3 Dynamic Conic Hedging

Some of the above sections dealt with one-step trees. Here we make the procedure dynamic and consider multi-step trees, with a rebalancing at each time-step as the main new feature. We start by explaining the procedure for a two-step trinomial tree in which we perform a dynamic conic delta hedge. In the next section, we move to a more realistic setting where a multi-step multinomial tree is deployed.

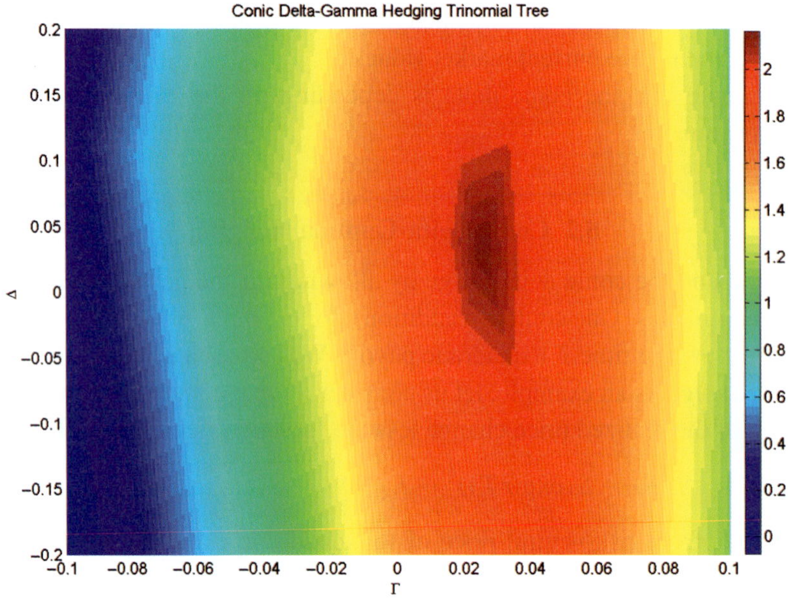

Figure 8.7 Conic delta-gamma hedging under a trinomial tree

8.3.1 Dynamic Conic Hedging in a Trinomial Tree

We consider the setting of Section 2.7.3 but now also calculate bid and ask prices in each of the nodes, as illustrated in Figure 8.8. The exact formula depends on the ordering of the payoffs in the final state. Let us work out the example for a particular setting, namely

$$f_{uu} \leq f_{um} \leq f_{mm} \leq f_{md} \leq f_{dd}.$$

To such a situation, a cdf as in Figure 8.9 is corresponding (on a one-step level).

One observes that the distorted probability to receive the up-state value is now $p_u^* = \Psi(p_u)$. With a probability $p_m^* = \Psi(p_u + p_m) - \Psi(p_u)$ one gets the middle-state value and finally the down-state value is received under distortion with a probability $p_d^* = 1 - \Psi(p_u + p_m)$. This leads to the bid price at the intermediate states:

$$\begin{aligned}
\text{bid} f_u &= \exp(-r\Delta t)\left(p_u^* f_{uu} + p_m^* f_{um} + p_d^* f_{mm}\right)\\
\text{bid} f_m &= \exp(-r\Delta t)\left(p_u^* f_{um} + p_m^* f_{mm} + p_d^* f_{md}\right)\\
\text{bid} f_d &= \exp(-r\Delta t)\left(p_u^* f_{mm} + p_m^* f_{md} + p_d^* f_{dd}\right).
\end{aligned}$$

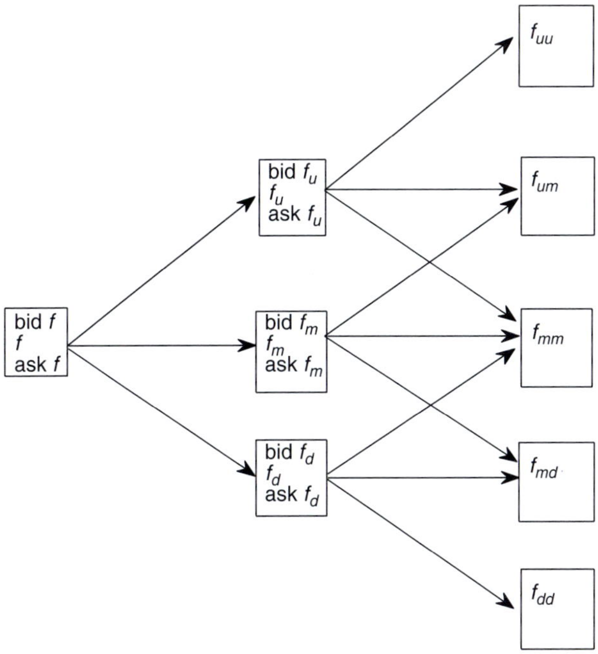

Figure 8.8 Two-step trinomial tree

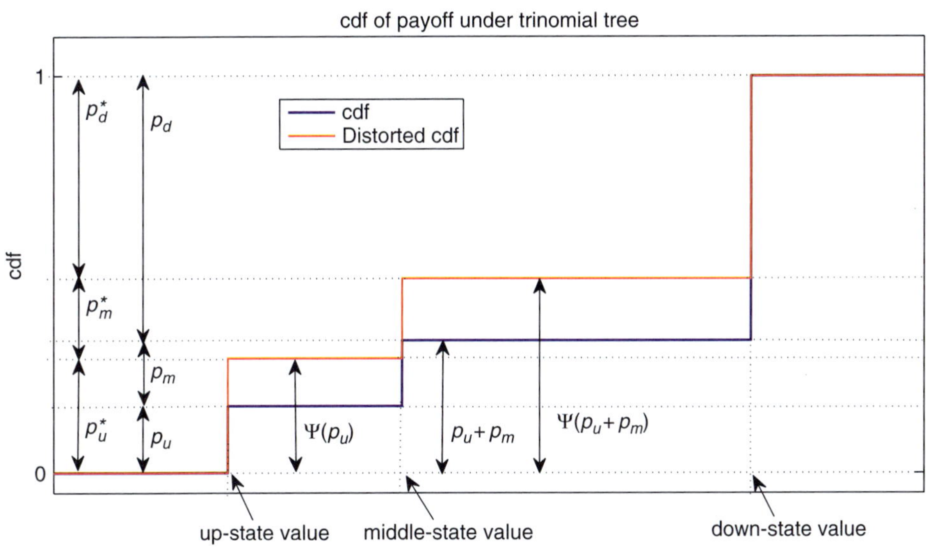

Figure 8.9 cdf and distorted cdf under a trinomial tree setting with $f_m \geq f_d \geq f_u$

Similarly, using that in this case $\text{bid} f_u \leq \text{bid} f_m \leq \text{bid} f_d$, the initial bid price is calculated as:

$$\text{bid} f = \exp(-r\Delta t)\left(p_u^* \text{bid} f_u + p_m^* \text{bid} f_m + p_d^* \text{bid} f_d\right).$$

Hence the final bid price is calculated as a bid price of the intermediate bid values. Ask prices can be calculated in an analogue way. Recall that the ask price is the negative of the bid price of the negative values/payouts.

Further, one can also calculate at the intermediate nodes the optimal conic bid delta hedge and the associated (maximized) bid values. The same is true for the conic ask delta hedge and the associated (minimized) ask values. Then using these optimized bid or ask values, one can calculate the conic bid and ask delta hedge and the corresponding optimal bid and ask prices at the initial state.

Example 8.1 *Let us set up a two-step trinomial tree with $T = 1/12$ and $\Delta t = T/2 = 1/24$. Assume that we have a non-dividend-paying stock with $S_0 = 100$ with a yearly volatility given by $\sigma = 0.2$. Assume further that $r = 0.01$. We will make use of the MINMAXVAR distortion and use a $\lambda = 0.125$ at each time-step.*

For this setting we have

$$d = 0.9313, \quad m = 0.9996, \quad u = 1.0728$$

and corresponding probabilities of moving respectively down, middle or up of $1/6$, $2/3$ and $1/6$. Let us price in this example a put with strike 105 and hence

$$f_{uu} = 0 \leq f_{um} = 0 \leq f_{mm} = 5.0883 \leq f_{md} = 11.9045 \leq f_{dd} = 18.2600.$$

As distorted probabilities we have:

$$\begin{aligned}
p_u^* &= \Psi(1/6) = 0.2257, \\
p_m^* &= \Psi(5/6) - \Psi(1/3) = 0.8820 - 0.2257 = 0.6563, \\
p_d^* &= 1 - \Psi(5/6) = 1 - 0.8820 = 0.1180.
\end{aligned}$$

From this we can calculate

$$\begin{aligned}
bid f_u &= \exp(-r\Delta t)\left(p_u^* f_{uu} + p_m^* f_{um} + p_d^* f_{mm}\right) = 0.5995, \\
bid f_m &= \exp(-r\Delta t)\left(p_u^* f_{um} + p_m^* f_{mm} + p_d^* f_{md}\right) = 4.7388, \\
bid f_d &= \exp(-r\Delta t)\left(p_u^* f_{mm} + p_m^* f_{md} + p_d^* f_{dd}\right) = 11.1102.
\end{aligned}$$

The risk-neutral prices are respectively, 0.8469, 5.3707 and 11.8219; the ask prices are 1.1468, 6.0206 and 12.5289. From these prices the initial prices can be calculated. The bid, for example, is calculated via the formula:

$$bid f = \exp(-r\Delta t)\left(p_u^* bid f_u + p_m^* bid f_m + p_d^* bid f_d\right) = 4.5544.$$

Table 8.2 *Conic delta hedging under trinomial tree (intermediate state).*

State	Risk-neutral	Conic delta hedged bid	Unhedged bid	Δ^{bid}_{conic} (hedge)
up	0.8469	0.6556	0.5995	0.3349
middle	5.3703	5.2923	4.7388	0.8418
down	11.8219	11.8219	11.1102	1.0000

Similarly, we get a risk-neutral price of 5.6896 and an ask price of 6.9116.

Next, we put on a conic delta hedge. We summarize the results for the intermediate states in Table 8.2.

Finally, for the initial state we obtain a conic delta hedged bid price of 5.5223, whereas the unhedged bid would be 4.5544; the risk-neutral price equals 5.6896. The optimal conic delta hedge for the bid in the initial state equals $\Delta^{bid}_{conic} =$ 0.7893.

8.3.2 Dynamic Conic Hedging in a Multinomial Tree

The focus of attention in the above binomial and trinomial tree models has been on complete or to-be-completed market approximations. Now we focus our attention on managing, evaluating and comparing hedge designs in the presence of residual risk essential to incomplete market contexts. We study multinomial approximations to one-dimensional processes that have exposure to many moves in an instant. In particular, we proceed to use the multinomial approximation constructions of Section 2.7.4.

We take the local motion to be multinomial and couple this by storing one-step-ahead value processes on a grid from which one may interpolate the one-step-ahead values needed. More precisely, we will first construct a (non-uniform) grid for the stock price dimension around the current value. Further, we will identify the time points at which a re-hedging and revaluation will be performed. The most convenient way to do this is to take equidistant time steps. For example, if we study a one-year-to-maturity derivative, we can take weekly time-steps. Then, we start our analysis and calculate the payoff values at maturity for each of the stock values in our grid. Next, we move one step backward in time and calculate, for each of the stock values in our grid, the bid and ask prices (with and without conic hedging). We do this by assuming that local motion is multinomial. That is, given we are at a given stock price level, we assume that the stock can move only to a finite number of stock values at maturity. The exact values are determined via the multinomial model in place. In order to calculate the bid and ask prices we are interested in, we need to have the payoff values in these final stock values. We calculate these by interpolation on our stock price grid, for which we just calculated the payoff values. Now, we are in a position to apply the techniques of the above sections but in a multinomial setting. Once we have calculated the bid and ask prices, we can move again one step backward in our time grid. At this time point we repeat the procedure and calculate for each stock price in our stock price grid the bid and ask values assuming again the same local multinomial motion. The difference with the previous step is that now the bid prices are calculated using the one-step-ahead bid prices (which we just calculated) – and the same for the ask prices. For example, the conic delta hedged bid price at this particular stock level is calculated using all the one-step-ahead conic delta hedged bid prices in the possible stock values our local multinomial model can jump to. These one-step-ahead conic delta hedged bid prices at our multinomial grid are derived using interpolation on the one-step-ahead conic delta hedged bid prices of our stock price grid (which were calculated just before). We repeat this procedure again and move one further time step backward. We keep repeating this until we have calculated at our initial time point the relevant values for the whole stock price grid. The final result is now a variety of matrices with bid and ask prices corresponding to a particular time point and stock value.

We illustrate this procedure on a call spread, with a one-year maturity $T = 1$ and a weekly rebalancing (say $N = 50$). We assume $S_0 = 100$, $r = 0.02$ and will investigate the conic hedging of a 110–120 call spread with payoff:

$$(S_T - 110)^+ - (S_T - 120)^+.$$

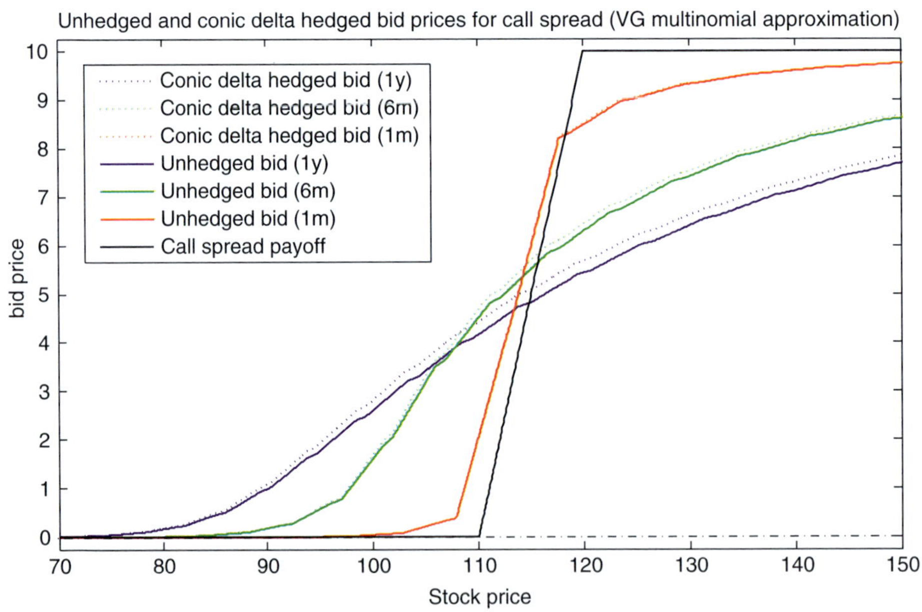

Figure 8.10 Unhedged and conic delta hedged bid prices for call spread

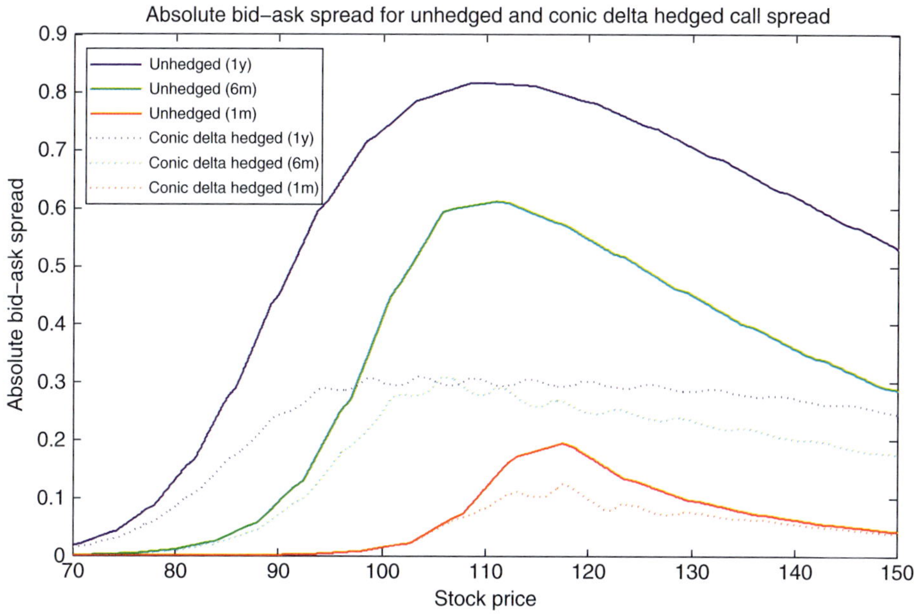

Figure 8.11 Bid–ask spreads for hedged and unhedged call spread

As an example, we again use the stylized annualized parameters $\sigma = 0.2$, $\nu = 0.75$ and $\theta = -0.3$ (as in Example 2.3). We create an 800-point (non-uniform) stock price grid with stock values ranging from 50 to 200. We take the same multinomial approximation as in Example 2.3, where we have 21 different jump sizes (including no jump). We use the MINMAXVAR distortion with stress level $\lambda = 0.015$ in each time step.

We first calculate the unhedged and the conic delta hedged bid and ask values. We show in Figure 8.10 the unhedged and conic delta hedged bid curves at the initial time point (one year to maturity), halfway (or after 25 time steps) and four time steps (i.e. about one month) before maturity.

We further report on the bid–ask spread in Figure 8.11. The unhedged ATM spread is reduced by about 50% with the dynamic delta hedge that maximizes the bid price and minimizes the ask price.

9

Hedging Insurance Contracts

In this chapter we illustrate the conic finance theory in the setting of equity-linked insurance contracts. We elaborate on how to improve the risk profile and risk-manage such types of contracts (see also Carr et al. (2016)).

Insurance books often combine exposures to financial and non-financial risks, with the financial component often being quite complex and involving path-dependent aspects. Yet one anticipates that movements in security prices over the near term can have considerable effects on the present value of payouts associated with the insurance liabilities. Taking positions in the options markets of related securities may constitute a good hedge for the insurance book. The purpose here is to present a detailed methodology for constructing such option hedges. The purpose of the hedge is not that of replication or risk elimination, as this is not expected to be possible, but one of improving the overall quality of the position.

Implementation of the methodology requires an ability to simulate the paths, at a regular frequency, of relevant underliers and the associated present value of the related insurance cash flows. We introduce as hedging assets positions in the underlying stocks and out of the money options on these stocks. The residual position that we investigate is given by earnings on the hedge positions together with the present value of the insurance contract's cash flows. The hedge positions are sought in a way that makes the package more diversified. Diversity is measured by the gap between the ask price of the package and the sum of the ask prices of the components of the package. We hence face an optimization problem in the number of all options on all underliers plus the number of underliers for the stock positions. The hedge is designed to numerically solve this optimization problem.

We exemplify the methodology on Guaranteed Minimum Withdrawal Benefit Variable Annuities (GMWBVA). When such deals are struck, or more generally whenever these deals are on the books, they are exposed to market risk. Using a particular model, one can try to estimate this risk and, for example, estimate what the marked-to-market profit or loss would be if the underlying stock or index were to move. Knowing this, one could try to hedge away these market uncertainties by taking positions in derivatives that try to offset these changes. The issuer of these contracts then just carries the residual risk, i.e. the insurance risk. How to hedge the equity risk is a non-trivial question. On the one hand, these insurance contracts have a very long (often unknown) maturity; for example, they may terminate at the death of the contract buyer. On the other hand, liquid derivative instruments, like vanilla calls and puts, are only available on relative short maturities. Therefore, we are going to put on short-term semi-static hedges, in the sense that we will try to estimate the equity risk on a relatively short horizon and hedge this risk. After the hedge expires, a similar hedge will need to be set up again with a relatively new small horizon. We then move forward until the contract terminates or the equity risk has been reduced to such an extent that we are no longer really bothered about it.

We first explain the GMWBVA contract. We then show how to calculate the related equity risk and how to hedge it. The underlying idea will be that we will look for an equity derivatives position that, in combination with the contract, has a minimum ask price. Hence we will add derivatives to our insurance contracts in amounts that will maximize diversity (read minimize the ask price).

9.1 The Guaranteed Minimum Withdrawal Benefit Variable Annuities Contract

The GMWBVA works as follows. An investor pays into an account a certain sum, say EUR 100,000. Typically he does that when he comes close to his retirement age; say when he is 60. Typically the amount is invested in an underlying (stock) index. The contract will provide the buyer with a certain monthly withdrawal that is often considered as an extra income to compensate the reduction in income after retirement.

Together with the actual account, we also keep track of a so-called *base*. The base initially is set equal to the initial amount. The value of the account and base will vary over time and their time t values will be denoted by, respectively, A_t and Z_t; we denote the price at time t of the underlying index by S_t.

On a regular basis, say monthly, the account value and base value are recalculated. The base is reset according to the contract terms and conditions. The investor has to pay some management fees. Further, and also according to the contract, certain periodic withdrawals are paid out. In addition, we have the market dynamics at play moving the underlying equity value. Finally, the contract can be terminated due to the death of the contract buyer, and this often goes together with a final redemption.

Next, we specify the cash flows and their timing in more detail for our specific example product. We start by specifying when the contract terminates. The account value process and its associated base evolves until the random time $\tau > 0$ that is the minimum of either the time of death of the account holder or the time at which the account value hits zero. Positive account values at death are returned to the account holder's estate and all premium payments and withdrawals stop. If the account value hits zero before death, the account holder receives a percentage of the base to death (his usual withdrawals) and all fee payments stop. We typically have a management fee κ based on the account value and a fee y relative to the level of the base. We denote the withdrawal rate on the base by z.

Assuming monthly withdrawals and rebalancing, this means that on the one hand at the end of the first month, the account is reduced by

- the management fee on account: $\kappa A_0 h$
- the management fee on base: $y Z_0 h$
- a withdrawal: $z Z_0 h$

where $h = 1/12$, a month.

On the other hand, the account also reflects the returns to investment in the equity exposure. This can be modelled with some estimated market model and we employ here in our illustration a VG model:

$$S_t = S_0 \exp((r - q + \omega)t + X_t), \qquad t \geq 0,$$

where as usual r equals the risk-free rate, q is the dividend yield and ω is given as in Section 2.5.3. We hence have that our initial investment of A_0 after one month would show an absolute return of

$$A_0(\exp((r - q + \omega)h + X_h) - 1).$$

Hence in total our account after one month, including the equity return and paying out fees and withdrawal, would equal

$$\begin{aligned} A_h &= A_0 - \kappa A_0 h - (z + y) Z_0 h + A_0(\exp((r - q + \omega)h + X_h) - 1) \\ &= A_0 \exp((r - q + \omega)h + X_h) - \kappa A_0 h - (z + y) Z_0 h. \end{aligned}$$

The base is also reset at the end of each month. In our example we assume a monthly increase of the base at a rate of η per annum for the first 10 years, say, and zero thereafter. Furthermore, one always compares the base with the account and takes the maximum:

$$Z_h = \max(Z_0(1 + \eta h), A_h).$$

We now also check whether $A_h \leq 0$ or whether the account holder died. In the latter case, the account value (if positive) is just paid out and the contract stops. In the former case, no fees will be paid out in future periods and the contract holder is paid out his monthly withdrawal (z) on the base (at the moment the account hit zero) till death. There is then no longer any equity exposure and one can actually then calculate what the expected present value is of these withdrawals using a particular lifetime distribution (see below).

For further time steps conditioned on the fact that the account didn't hit zero earlier and that the contract buyer did not die, we have similar kinds of recalculations:

$$A_{t+h} = A_t \exp((r - q + \omega)h + (X_{t+h} - X_t)) - \kappa A_t h - (z + y)Z_t h$$

and

$$Z_{t+h} = \max(Z_t(1 + \eta h \mathbf{1}(t + h \leq 10)), A_{t+h}).$$

Note the small difference in the recalculation of the base, where we now explicitly check whether we are in a period where the base is still increased ($t + h \leq 10$) or not. Recall that $\mathbf{1}(A) = 1$ if A is true and zero otherwise.

Let the complementary distribution function for the time of death of the contract holder be

$$G(t) = P(\text{contract holder lives longer than age } t).$$

If the account value hits zero and death has not occurred, then for an account holder who had an age α when entering the contract at time zero (given of course he was alive!), the present value at time t of the withdrawal payouts (with withdrawal rate z) on the base until death is

$$V(t) = Z(t)z \int_0^\infty \exp(-rs)G(t + \alpha + s)\mathrm{d}s / G(\alpha).$$

For example, we could use a Weibull distribution for life expectancy, then

$$G(t) = \exp\left(-(t/c)^a\right), \qquad t \geq 0,$$

for certain parameters $a > 0$ and $c > 0$. The mean and variance of the Weibull distributed random variable X are given by:

$$E[X] = c\Gamma\left(1 + \frac{1}{a}\right) \text{ and } Var[X] = c^2\left(\Gamma\left(1 + \frac{2}{a}\right) - \left(\Gamma\left(1 + \frac{1}{a}\right)\right)^2\right).$$

For a given average life expectancy and a given standard deviation, one can, by employing these equations via a numerical search, obtain the matching parameters c and a.

For example, for a mean of 70 and a standard deviation of 10, we get $c = 74.1885$ and $a = 8.3245$. The corresponding $G(t)$ is plotted in Figure 9.1. Further for $Z(t) = 100$ and $z = 5\%$, a picture of $V(t)$ is given in Figure 9.2. If the account hit zero after 10 years and the account holder is still alive, the present value

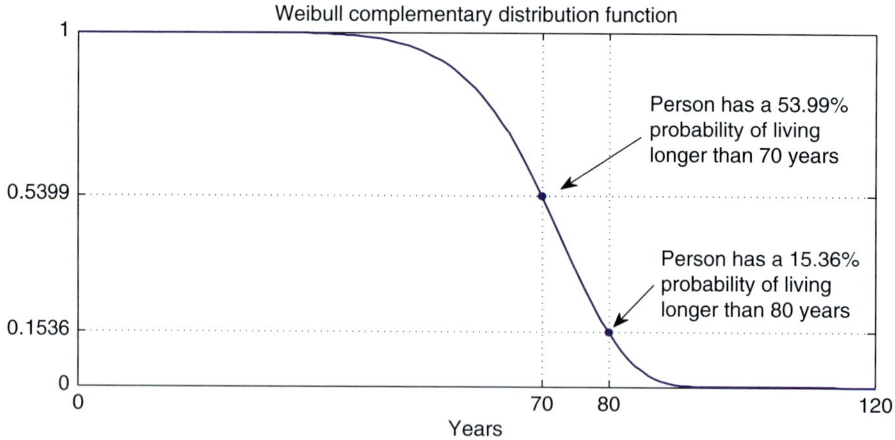

Figure 9.1 Weibull complementary distribution function with mean of 70 and a variance of 10

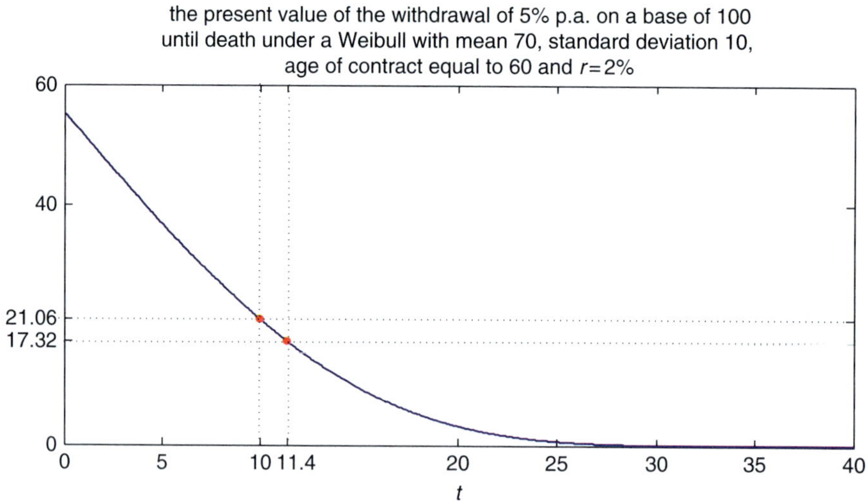

Figure 9.2 The present value of the expected withdrawals under Weibull with mean of 70 and a variance of 10

of the expected withdrawals until his death equals $V(10) = 21.06$. If the account hits zero after 11 years and 5 months, the present value of the expected withdrawals until his death equals $V(11.4167) = 17.32$.

In Figure 9.3, two examples of scenarios are depicted. In the first scenario the account holder dies at the age of 62 and 11 months, i.e. 2 years and 11 months after initiating the GMWBVA contract. At that moment his account value is 102.29 and this amount is then paid out to the holder's estate. In the second scenario the account's value hits zero after 11 years and 5 months. The account holder is then still alive. From then on he receives $z = 5\%$ p.a. of the base which is then at 182.14

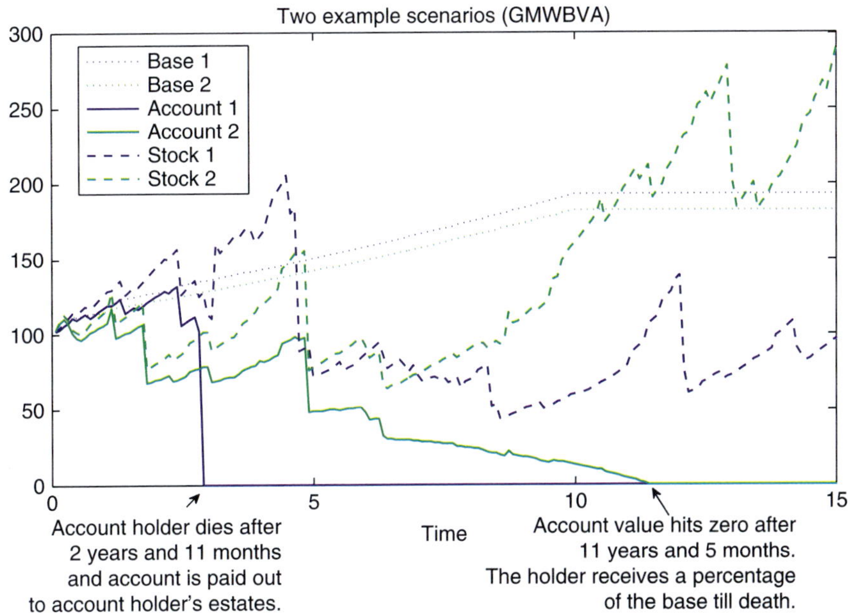

Figure 9.3 Two examples of scenarios

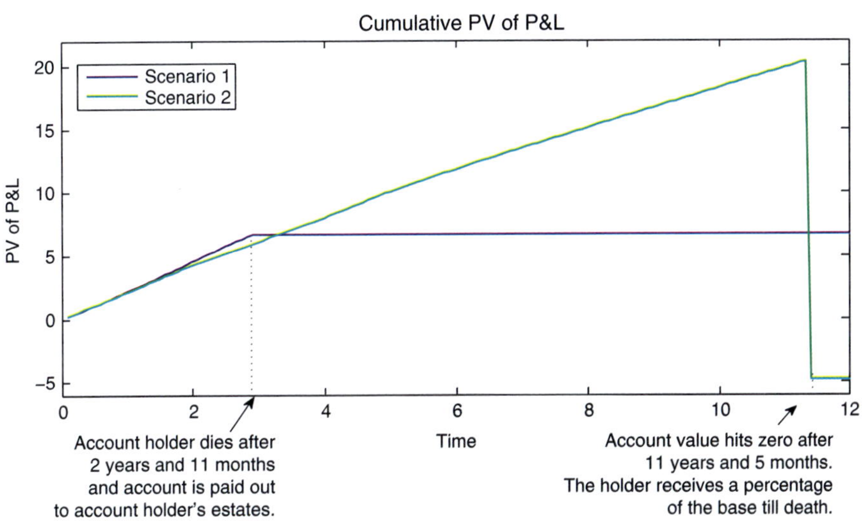

Figure 9.4 The present value (PV) of the cumulative P&L for the two examples of scenarios

till he dies. For the insurance company this is a bad scenario, since it has to come up with these payments. The present value of these further payments till the account holder's death is 31.55 ($= 17.32 \times 182.14/100$).

In Figure 9.4, the present value (PV) of the cumulative P&L for the insurance company is shown. The insurance company accumulates fees either until the account holder dies (scenario 1) or until the account value hits zero (scenario 2). In the latter, the company has to pay the monthly withdrawals until the account holder's death. In the picture this is shown as an one-time drop of about $25.11 (= \exp(-r(11+5/12)) * 31.55)$ in the accumulated present value of the P&L after 11 years and 5 months.

9.2 Conic Risk Management of Equity-Linked Insurance Contracts

For a single account, we simulate a large number N, say a million, of paths of length 480 (representing 40 years of monthly steps) for the underlying equity, the account value, the level of the base and the cash flows paid. The account holders received as a lump sum the expected present value of $z = 5\%$ of the base from the time the account value goes to zero to death, when the former was earlier.

From these paths one can calculate the P&L for the insurance company. It equals the present value of all payments, the fee payments (positive) and the lump sum payment (negative) through time. In our setting, we store these aggregated present values in the column vector C of length one million. Figure 9.5 presents a histogram of this P&L.

Assume a fixed short maturity of say $T = 0.5$ (6 months). We perform two conic hedging exercises. The first is a conic Δ-hedge where only a forward position is

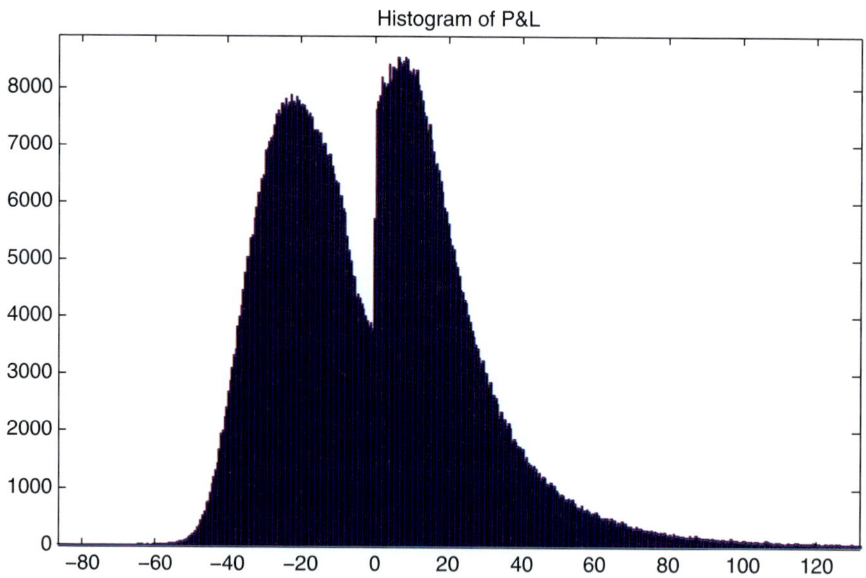

Figure 9.5 Histogram of P&L. A positive value means a profit for the insurance company

added to our contract. The second hedge is a conic option hedge and we assume we have at our disposal a series of OTM options maturing at time T.

In our example, we assume there are in total six options available. Along with the underlying stock itself it gives us a hedging asset space of $n = 7$ hedging assets. More precisely, we have four puts with strikes 70, 80, 90 and 100 and two calls with strikes 110 and 120. We denote the strikes by $K_i = 70 + 10(i-1), i = 1, \ldots, n-1$ and the current value of these vanilla options by $w_i, i = 1, \ldots, n-1$.

The main idea is to combine our insurance contract with (zero-cost) positions in the hedging assets, such that the portfolio is more diversified, i.e. hedged. More precisely, we will work out an ask price minimization. We only allow in this illustration long positions. We hence perform the conic hedging methodology of Chapter 8 but now using either 1 or 7 hedge instruments.

Diversification can be measured by the distance of the ask price of the package to the sum of the ask prices of the different components of the package. The former being always below the latter. The wider the gap the more diversification there is. We hence look for the package with the lowest ask price, i.e. with the widest gap, and hence the most diversified one.

In concrete terms, we are looking for a vector $x = (x_0, x_1, \ldots, x_{n-1})$ of size n, where x_0 represents the position with exposure to the underlying asset/forward and $x_i, i = 1, \ldots, n-1$ the respectively exposures to the options, if any. More precisely, we are looking for a vector x such that the price of the package consists of

$$Y = x_0(S_T - \exp(rT)S_0),$$

for the pure conic delta-hedge exercise, or for the option hedge

$$Y = x_0(S_T - \exp(rT)S_0) + \sum_{i=1}^{4} x_i \left((K_i - S_T)^+ - \exp(rT)w_i\right)$$
$$+ \sum_{i=5}^{6} x_i \left((S_T - K_i)^+ - \exp(rT)w_i\right),$$

combined with the initial insurance contract has the lowest ask price. Note that the current risk-neutral price of Y is equal to zero. We are thus adding to our insurance contract a zero-cost position.

Since we have generated a path space, we can calculate the ask price of any package. We evaluate for each stock path the payoff of Y. Let us store the discounted value of these payoffs in a column vector H of size one million. For the insurance company, the total P&L for the jth path in our path space equals

$$\text{P\&L}_j = H_j + C_j, \qquad j = 1, \ldots, N.$$

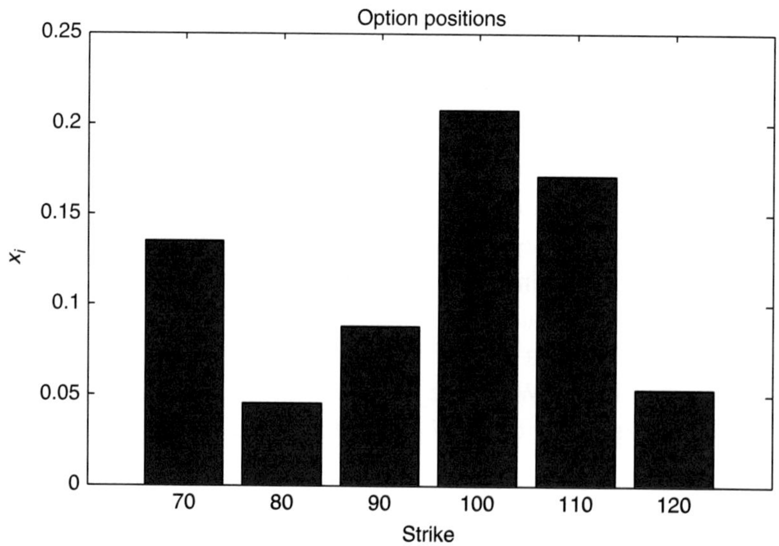

Figure 9.6 Option positions

We can apply the same pricing techniques of Section 5.3 and in particular Formula (5.5) on this P&L vector, and we can easily calculate the conic ask price under our favourite distortion function. We use the MINMAXVAR distortion with a stress parameter $\lambda = 0.75$. We then optimize by doing a search over the position vector x, looking for the lowest ask price of the package.

For the delta-hedge we obtain an optimal, i.e. minimal, ask price for a stock position of $x_0 = -0.091$. For the option hedge, performing a constrained optimization enforcing $x_i \geq 0$, $i = 1, \ldots, n-1$, we obtain an optimal for $x_0 = 0.095$ and option positions as graphed in Figure 9.6.

To illustrate the purpose of the hedge we provide an estimate for the expected P&L conditioned on the stock level after six months. We estimate this by performing a kernel density estimation. We observe in Figure 9.7 that the (left) tail-risk has been reduced.

Further, we calculate the standard deviation and skewness of the P&L of the unhedged and hedged cash flows. The values are shown in Table 9.1. The average P&L in both cases equals 0.23, so the deal is fairly priced with a bit of margin for the insurance company. Hedging reduces the variance of the P&L and delivers a more positively skewed P&L. Note that we have only put on one single hedge at time zero, which matures after six months. In practice, this methodology would be dynamically repeated. After the first hedge has matured, or even before it matures, a new similar hedge could be set up using short term options then available in the market. It is anticipated that performing such a dynamic version of this conic hedging would further reduce the variance of the P&L.

Table 9.1 *Standard deviation and skewness of hedged and unhedged P&L.*

Hedge	Standard deviation	Skewness
Unhedged	27.32	1.61
Delta-hedged	27.04	1.63
Option hedged	26.96	1.67

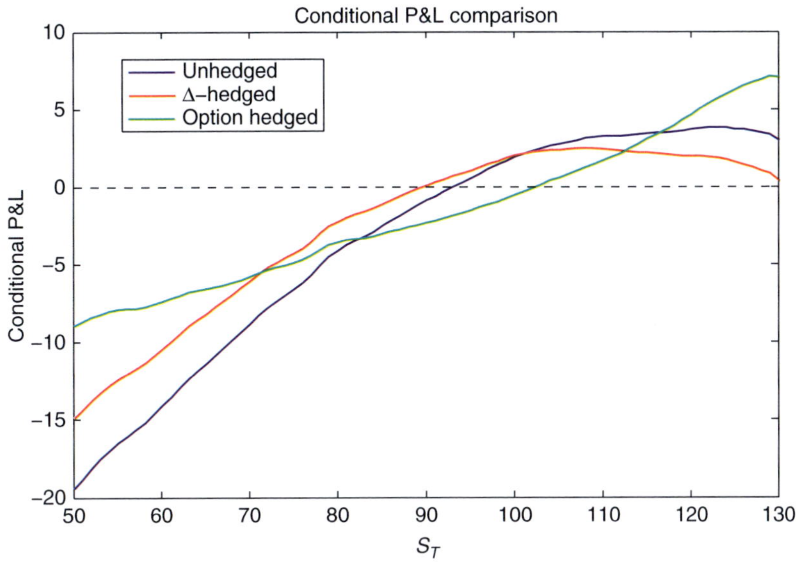

Figure 9.7 Conditional PV kernel estimate

We finally remark that often complex insurance risks typically have multiple exposures. The above methodology can be employed in a similar fashion on multiple underliers.

10

Option Positioning

In this chapter we focus on the problem of taking a good position in options on an underlying asset at a fixed maturity, T say, given an assessment of the physical measure P and the risk-neutral pricing measure Q using conic finance as a tool.

10.1 Option Positioning Using Conservative Conic Valuation

Given traded strikes for out of the money calls, denoted by $K_i, i = 1, \ldots, N$ and out of the money puts $\tilde{K}_j, j = 1, \ldots, M$, one seeks a stock position α and out of the money call and put positions, $\beta_i, i = 1, \ldots, N$ and $\gamma_j, j = 1, \ldots, M$ respectively, and a bond face value δ to get at time T the cash flow

$$c(S_T) = \delta + \alpha S_T + \sum_{i=1}^{N} \beta_i (S_T - K_i)^+ + \sum_{j=1}^{M} \gamma_j (\tilde{K}_j - S_T)^+.$$

For a large number of strikes with a low inter-spacing of strikes, it is well known that essentially all twice-differentiable functions may be approximated in this way. The abstract problem may then be formulated as seeking the positions $\alpha, \beta_i, \gamma_j, \delta$ or as seeking the function $c(s)$ that may subsequently be replicated by a bond, stock and option positions.

Of course, if one can access the function $c(s)$ it is clear that one may also access $kc(s)$ for any $k > 0$. In principle, one may then split the problem of selecting the function from that of selecting the scale.

Generally, market participants are interested in what to trade and not in how much to trade, as the latter is determined by the effects of market impact on executing trades. One is also often interested in capping the maximum loss associated with positions, as unlimited loss exposures are not tenable positions. This may be accomplished by focusing attention on non-negative cash flows for which the maximum loss is the upfront premium. We therefore focus for simplicity here on cases where $c(s) \geq 0$.

The functional problem of finding a non-negative cash flow subject to a budget constraint has been solved for the objective of expected utility maximization by a number of authors. The use of concave utility functions and the budget constraint renders a problem well posed with a good mathematical solution. However, both the use of a concave utility and the budget constraint are essentially artificial devices for attaining a well posed problem.

For the range of cash flows being contemplated or feasible, for many participants the marginal value or utility of a dollar remains a dollar and the modelling of the rate at which this is supposed to fall is hence an artificial exercise. The scale of the trade will be determined by other market considerations and one need not rely on essentially concocted concave utility functions with no foundations in the reality of the situation. The budget constraint is equally artificial as one is definitely willing to put on good trades that cost somewhat more or somewhat less. There is in effect no such budget constraint for most market participants. The budget constraint is just the cost of the trade that in principle can take on any real value in some unspecified interval, and one may consider moving out of this interval if the trade is good enough.

Maybe any attempt at formulating a well posed mathematical problem with a unique solution is bound to be artificial and hence without foundation or support in its execution or adoption. Trying to get the problem to be well posed with a unique mathematical answer is therefore not our focus. Yet a mathematical analysis of the problem can be helpful in guiding decisions on derivative positioning. Towards this end we formulate an objective function delivering an alternative conservative valuation for the cash flow being contemplated for access that one may compare with the cost at which it may be attained in the market. The good or admissible trades are those with a conservative valuation above the market cost. We seek to identify the trades with this property and to understand the sources for the higher conservative valuation, leaving it to the discretion of decision makers as to which of the many such trades, if any, they would like to put on. The ratio of the alternative conservative valuation to the market cost may be viewed as a favourable premium or impact factor, and one could backtest the strategy of putting on all trades identified with a premium crossing a threshold.

The alternative conservative valuation shares some properties with the market valuation. As we are not interested in focusing attention on determining scale issues, we take the alternative conservative valuation to scale with cash flows as do the market valuations (compare with Equation (4.3)). Further, as adding or subtracting constants from cash flows does nothing to the risk profile, we agree with the market in valuing all constant cash flows at the value of the underlying constant. Furthermore when added to a cash flow they enhance value by the same constant (compare with Equation (4.1)). Our model for an alternative conservative valuation is that for the bid price which is, as seen before, structurally the infimum of a set of candidate valuations conducted as expectations with respect to probability measures equivalent to a base or reference probability (compare with Equation (4.5)). We recall that, if the bid price is further taken to depend on just the probability distribution of the cash flows accessed and in addition one asks for additivity for the bid prices of comonotone components, then the bid price may be presented as an expectation under a concave distortion of the original cash flow distribution function (see Equation (4.8)).

Denote by $F_X(x)$, the physical (under P) distribution function of the cash flow X and denote the pricing (under Q) distribution function by $G_X(x)$. We therefore model the alternative conservative valuation $V(X)$ for a cash flow by

$$V(X) = \exp(-rT) \int_{-\infty}^{+\infty} x \mathrm{d}\Psi(F_X(x)), \tag{10.1}$$

for some fixed concave distortion function $\Psi(u), 0 \leq u \leq 1$. The price or market cost of the cash flow X on the other hand is

$$W(X) = \exp(-rT) \int_{-\infty}^{+\infty} x \mathrm{d}G_X(x) = \exp(-rT) E_Q[X]. \tag{10.2}$$

A knowledge of the physical and risk-neutral measures then gives us access to both the alternative conservative valuation $V(X)$ and the market price $W(X)$. Since we are basically comparing $V(X)$ with $W(X)$, the included discounting does not change the conclusions and one can actually leave it out and work on the forward prices.

Note further that we are applying here the distortion to the physical measure to obtain a conservative value V and take the market cost of the trade under consideration as given by W. Later we allow for an inflation or deflation of the market price W with a factor to let investors adjust the cost according to their own beliefs and skills.

10.2 Consequences of Valuing Just the Cash Flow Distribution Function

First observe that many functions of the underlying S_T at our time horizon can have the same distribution function. Indeed, let $y(s)$ be an arbitrary function of the underlying taking values on the real line, denote by $F_Y(y)$ the physical distribution function of $y(S_T)$ and suppose we aim for a non-negative cash flow with distribution function $H(c), 0 \leq c \leq 1$ (with inverse $H^{[-1]}(u)$), then both

$$c_1(S_T) = H^{[-1]}(F_Y(y(S_T))) \text{ and } c_2(S_T) = H^{[-1]}(1 - F_Y(y(S_T)))$$

have the same distribution function $H(c)$.

Indeed, first note that for any random variable X with cdf $F_X(x)$, $F_X(X)$ is uniformly distributed over the unit interval. Further observe that for such a uniformly distributed random variable U, we have that $1 - U$ is also uniformly distributed over the unit interval: $P(U \leq u) = P(1 - U \leq u) = u$. Therefore,

$$P(c_1(S_T) \leq c) = P(H^{[-1]}(U) \leq c) = P(U \leq H(c)) = H(c)$$

and

$$P(c_2(S_T) \leq c) = P(H^{[-1]}(1 - U)) \leq c) = P(1 - U \leq H(c)) = H(c).$$

The first cash flow is basically going long in the function $y(S_T)$; the second cash flow goes short in $y(S_T)$. Indeed, if $y(S_T)$ increases, then $F_Y(y(S_T))$ increases and hence $H^{[-1]}(F_Y(y(S_T))) = c_1(S_T)$ increases. Similarly, if $y(S_T)$ increases, then $1 - F_Y(y(S_T))$ decreases and hence $H^{[-1]}(1 - F_Y(y(S_T))) = c_2(S_T)$ decreases.

The most simple example is just $y(s) = s$ and we have two trades that either go long or short the underlier.

For a given specific trade function $y(s)$, the first question we address is that of constructing distribution functions $H(c)$ that maximize conservative valuation $V(c_1(S_T))$ (compare with bid price) subject to the constraint that $c_1(S)$ is feasible for a given cost $W(c_1(S))$. This is the problem of finding the best cash flow going long in $y(S_T)$ for a fixed cost $W(c_1(S))$. Similarly, we also solve the problem of maximizing the conservative price (compare with bid price) $V(c_2(S_T))$ subject to the constraint that $c_2(S)$ is feasible for a given cost $W(c_2(S))$. This is the problem of finding the best cash flow going short in $y(S_T)$ for a fixed cost $W(c_2(S))$.

The solution produces two frontiers that we may denote by $V_1(W)$ and $V_2(W)$ for each specific trade function. When acceptable trades are defined by a positive value for $V(W) - W$ then the best acceptable trade is the one maximizing $V(W) - W$. When it exists, it corresponds to a unit tangent to the frontier $V(W)$ having a positive intercept.

A bit more generally, we can also maximize $V(W) - \alpha W$ for a certain α, leaving us the possibility to inflate or deflate the market price with a factor $\alpha > 0$. When the optimal solution exists it corresponds to a tangent with slope α to the frontier $V(W)$ having a positive intercept.

For the solution of the W-constrained optimal distribution function H, we may consider the case of $y(s) = s$ with physical probabilities $F_S(s)$ and pricing probabilities $G_S(s)$. For other functions one just applies this solution to the physical probability density $F_Y(s)$ and pricing probability $G_Y(s)$ for $y(S)$ inferred from those on S. We therefore focus attention on the identity function $y(s) = s$ and the two cash flow functions

$$c_1(S_T) = H^{[-1]}(F_S(S_T)) \text{ and } c_2(S_T) = H^{[-1]}(1 - F_S(S_T)). \tag{10.3}$$

Finding $c_1(S_T)$ and $c_2(S_T)$ is equivalent to finding $H(c)$ and hence to $H^{[-1]}(u)$. We denote the solutions by $H_1(c)$ (or equivalently $H_1^{[-1]}(u)$) and $H_2(c)$ (or equivalently $H_2^{[-1]}(u)$) respectively.

10.3 The Design of Monotone Cash Flows in the Underlying Asset Price

Recall that we have at our disposal estimates of the physical probabilities $F_S(s)$ and pricing probabilities $G_S(s)$.

The solution for the monotone increasing case is given by

$$H_1^{[-1]}(u) = \int_0^u \mathbf{1}\left[1 - \Psi(v) > \alpha(1 - G_S(F_S^{[-1]}(v)))\right] dv \tag{10.4}$$

and the solution for the monotone decreasing case is given by

$$H_2^{[-1]}(u) = \int_0^u \mathbf{1}\left[1 - \Psi(v) > \alpha G_S(F_S^{[-1]}(1 - v))\right] dv, \tag{10.5}$$

where $\mathbf{1}[A] = 1$ if A is true and is zero otherwise. The derivation can be found in Section 10.5.

For the purposes of illustrating the design of cash flows using a variety of trade functions we consider an annual maturity with physical distribution in the Variance Gamma class with parameters $\sigma = 0.15$, $\nu = 0.4$ and $\theta = -0.03$. The corresponding risk-neutral parameters are $\sigma = 0.2$, $\nu = 0.75$ and $\theta = -0.3$. The corresponding densities are shown in Figure 10.1; r and q are assumed to be zero for simplicity.

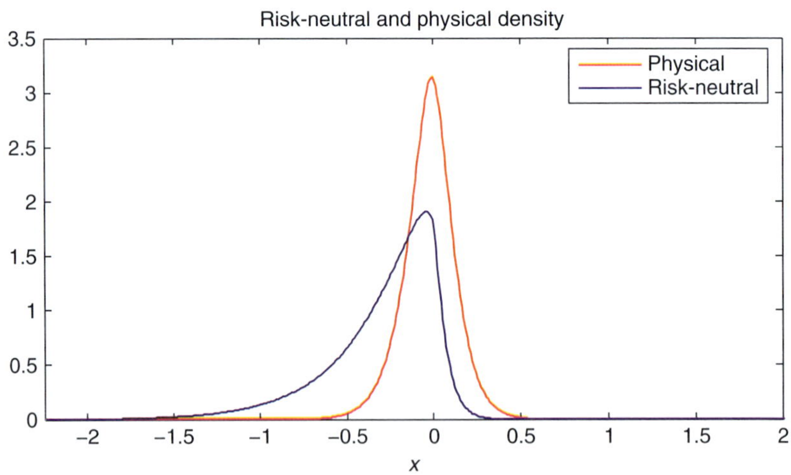

Figure 10.1 Risk-neutral and physical density

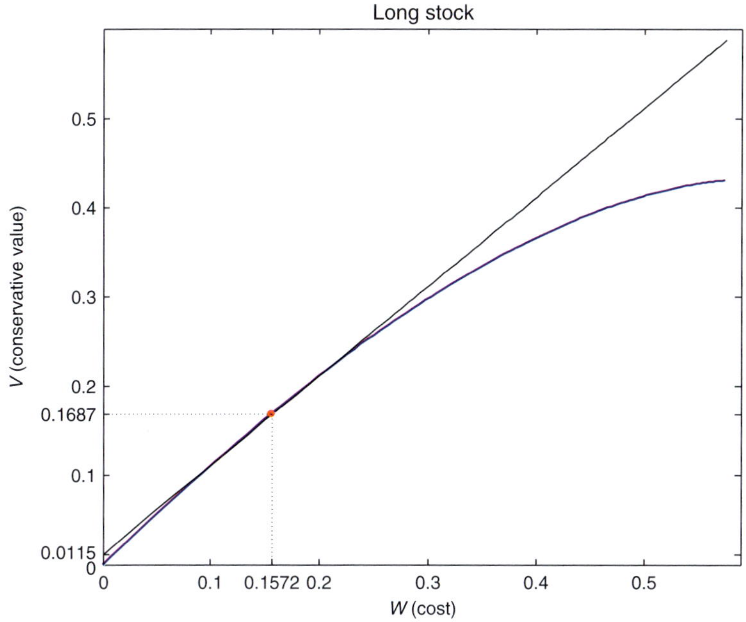

Figure 10.2 Value cost-efficient frontier of profit-maximizing trade going long the stock

Consider first claims that are long the underlying. In Figure 10.2, the *value cost-efficient frontier* is shown for a profit maximizing trade going long the stock. The figure is produced by calculating for a whole range of inflation values of α (we used αs in the range from 0.01 to 1.2 with steps of 0.01), first, the inverse distribution function $H_1^{[-1]}(u)$, using Equation (10.4), and then inferring from it $c_1(s)$, using Equation (10.3), and finally computing the corresponding values $V = V(c_1(S_T))$ and $W = W(c_1(S_T))$, using Equations (10.1) and (10.2).

We next focus on the particular values for $\alpha = 1$. For this value, the profit-maximizing trade has a cost of $W = 0.1572$ and a conservative value of $V = 0.1687$ with an expected profit of $V - W = 0.1687 - 0.1572 = 0.0115$. In the graph, we indeed see the unit tangent line at the corresponding values for V and W intersects the y-axis at the 0.0115 level.

Next, let us look at claims that are short the underlying. In Figure 10.3, the corresponding value cost-efficient frontier is shown now maximizing profit going short the stock. The figure is now produced by calculating for the same range of values of $0.01 \leq \alpha \leq 1.2$, first, the inverse distribution function $H_2^{[-1]}(u)$, using Equation (10.5), and then inferring from it $c_2(s)$, using Equation (10.3), and finally computing the corresponding values $V = V(c_2(S_T))$ and $W = W(c_2(S_T))$, using Equations (10.1) and (10.2).

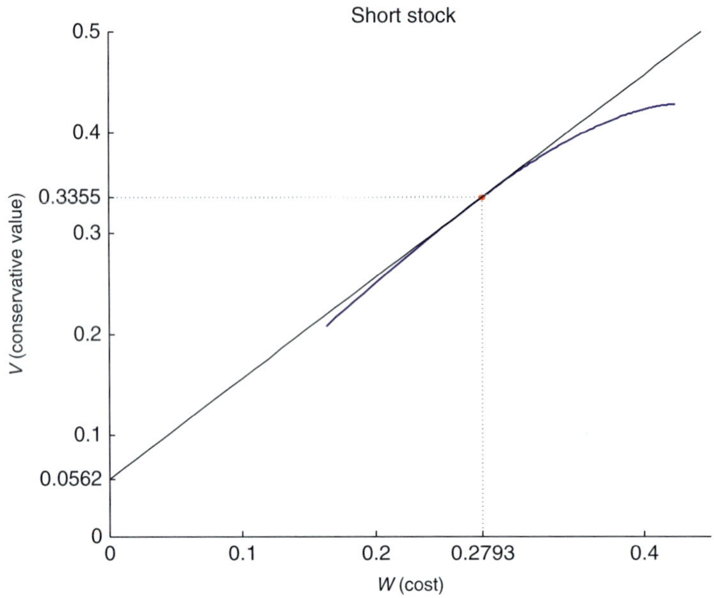

Figure 10.3 Value cost-efficient frontier of profit-maximizing trade going short the stock

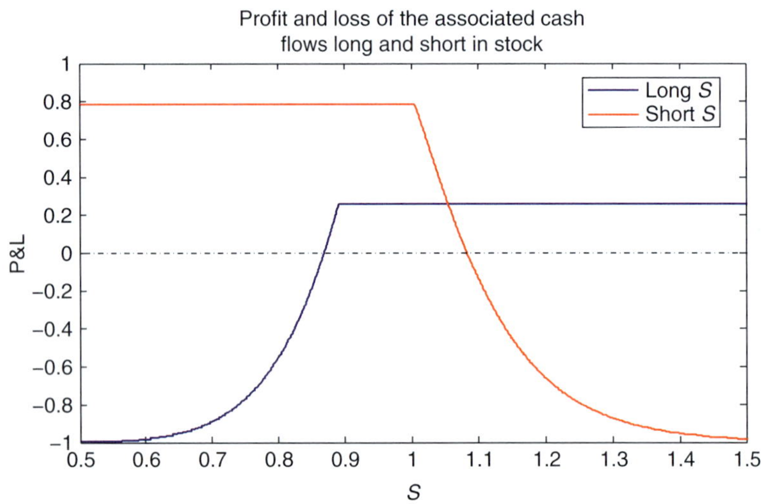

Figure 10.4 Profit and loss of the associated cash flows long and short in stock

For $\alpha = 1$, the profit-maximizing trade has a cost of $W = 0.2793$ and a conservative value of $V = 0.3355$ with an expected profit of $V - W = 0.3355 - 0.27493 = 0.0562$. The unit tangent line corresponding to the values for V and W intersects the y-axis at the 0.0562 level.

Figure 10.4 presents a graph of the profit and loss of the associated cash flows accessed. For comparison reasons, we have scaled both the long and the short trade such that it has a unit cost. We observe that the optimal long cash flow sells downside puts that are then bought back at lower strikes to keep the cash flow non-negative. It makes a profit if markets do not go down by more than about 13%. The optimal cash flow going short the stock sells at the money calls and then buys them back to keep the cash flows non-negative. The trade makes a profit if the market does not rise by more than around 8%.

10.4 The Design of Monotone Cash Flows in Other Trade Functions

As indicated before, any arbitrary trade function can be chosen to go long in or short in. We illustrate this by consider claims that are monotone increasing in the payoff to a strangle and a risk-reversal. The final case we present is for claims increasing in the ratio $f_S(s)/g_S(s)$, where $f_S(s)$ is physical density and $g_S(s)$ the pricing density. For such other functions $y(s)$ one just applies the above solution to the physical probability density $F_Y(s)$ and pricing probability $G_Y(s)$ for $Y = y(S_T)$ inferred from those on S_T. We employ the same VG setting as in the above section for the illustrations.

Optimal Positioning for Monotone Cash Flows in a Strangle

We first consider for $y(s)$ a strangle function. A strangle involves buying both a call option and a put option of the same underlying security. The owner of a long strangle makes a profit if the underlying price moves far enough away from the current price, either above or below. He may thus take a long strangle position if he thinks the underlying security will become highly volatile, but does not know which direction it is going to move in. Buying a strangle is basically going long volatility, so using this trade function can in that perspective be seen rather as looking for the optimal trade, having a view on either going long or short volatility.

In the illustration, our trade function in which we will go either long or short is:

$$y(S_T) = (0.9 - S_T)^+ + (S_T - 1.1)^+.$$

For $\alpha = 1$, there is no profitable, i.e. with $V - \alpha W > 0$, long position. There is an optimal short position; it has a conservative value of $V = 0.6947$ and a cost of $W = 0.2788$. As can be seen in Figure 10.6, the trade makes a profit if the underlying stock doesn't move with about 20%. A position short in a strangle indeed makes money if the stock remains rather flat and volatility low.

Optimal Positioning for Monotone Cash Flows in a Risk-Reversal

Next, we take for $y(s)$ a risk-reversal payoff function. A risk-reversal strategy consists of selling (or writing) an out-of-the-money put option and simultaneously buying an out-of-the-money call. Buying a risk-reversal is in general going long

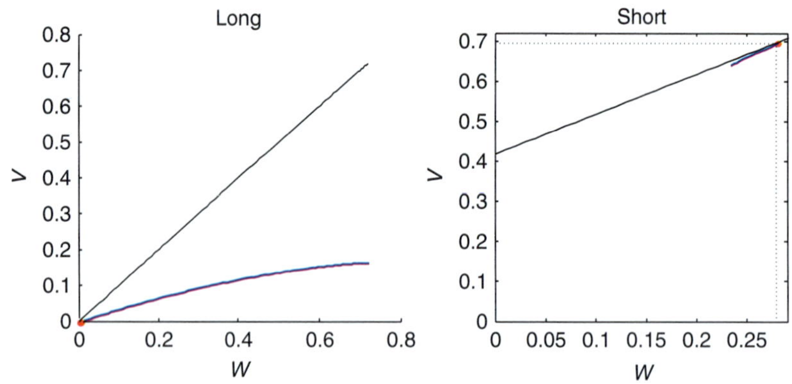

Figure 10.5 Value cost-efficient frontier of profit-maximizing trade going long the strangle and short a strangle

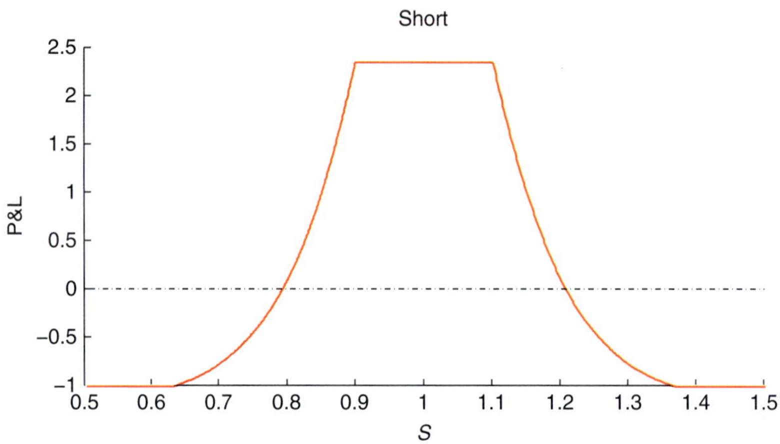

Figure 10.6 Profit and loss of the associated cash flows short in a strangle

the stock but in a more sophisticated way. Instead of going directly and linearly long on the stock, the buyer will buy an out-of-the-money call option, and simultaneously sell an out-of-the-money put option. Typically, one is using the money from the sale of the put option to purchase (partially) the call option.

In the illustration, our trade function in which we will go either long or short is:

$$y(S_T) = -(0.9 - S_T)^+ + (S_T - 1.1)^+.$$

For $\alpha = 1$, the optimal long position has a conservative value of $V = 0.1687$ and a cost of $W = 0.1572$. The optimal short position has a conservative value of $V = 0.1607$ and a cost of $W = 0.1402$. As can be seen in Figure 10.8, the long risk-reversal optimal trade makes money if the stock does not fall more than

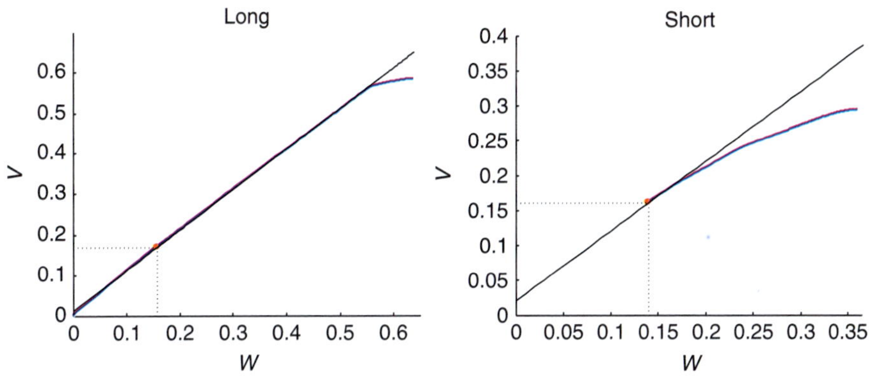

Figure 10.7 Value cost-efficient frontier of profit-maximizing trade going long and short a risk-reversal

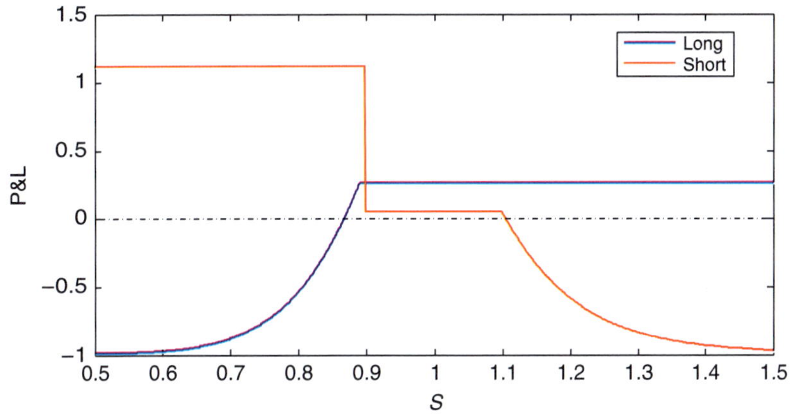

Figure 10.8 Profit and loss of the associated cash flows long and short in a risk-reversal

12% and is hence clearly also a position long in stock, as is the risk-reversal. The short risk-reversal optimal trade makes money if the stock does not rise more than 17% and is hence clearly also a position short in stock, again like the risk-reversal. The trades are just a bit more sophisticated, since the view they take, i.e. long/short risk-reversal, is a more sophisticated view than just being long/short stock.

Optimal Positioning for Monotone Cash Flows in the ratio of Stock Densities

Finally, we take for $y(s)$ a function in terms of the physical and pricing density, like in many solutions to the expected utility maximization problem. In the illustration, our trade function in which we will go either long or short is:

$$y(S_T) = \log(f_S(S_T)/g_S(S_T)).$$

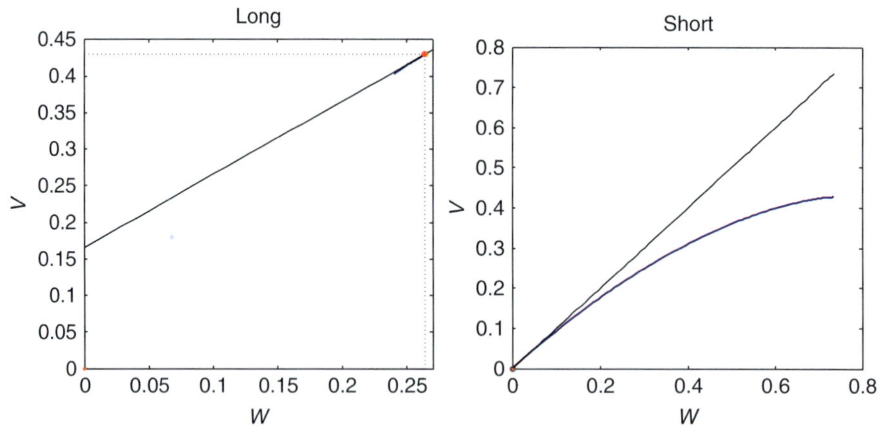

Figure 10.9 Value cost-efficient frontier of profit-maximizing trade going long and short the ratio of physical over pricing density

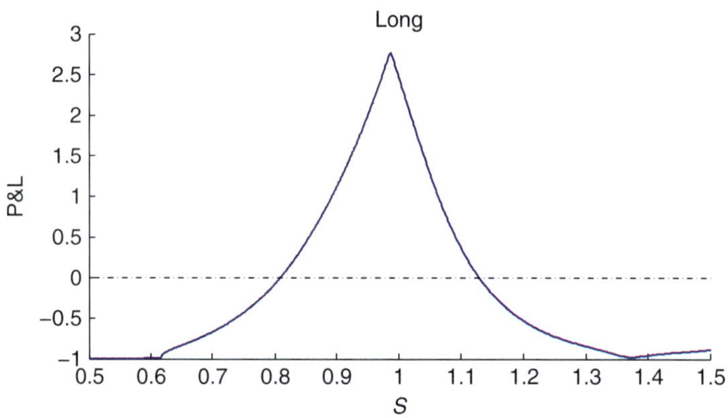

Figure 10.10 Profit and loss of the associated cash flows long in the ratio of physical over pricing density

For $\alpha = 1$, the optimal long position has a conservative value of $V = 0.4299$ and a cost of $W = 0.2642$, leading to an estimated conservative profit of 0.1657. The P&L of the optimal long trade is shown in Figure 10.10.

10.5 Appendix: Derivation of the Solution for Monotone Cash Flows

We only derive the solution for the monotone increasing case; the monotone decreasing case is completely analogously derived. Recall, we take $r = q = 0$ and we have at our disposal estimates of the physical probabilities $F_S(s)$ and pricing probabilities $G_S(s)$, or equivalently $f_S(s)$ and $g_S(s)$ as respectively physical and pricing density, and look for $c_1(S_T)$, which is equivalent with finding $H_1(c)$ and hence with $H_1^{[-1]}(u)$, since $c_1(S_T) = H^{[-1]}(F_S(S_T))$.

The constraint is given by

$$\begin{aligned} W = W(c_1(S_T)) &= E_Q[c_1(S_T)] \\ &= \int_0^{+\infty} c_1(s)\mathrm{d}G_S(s) \\ &= \int_0^{+\infty} H^{[-1]}(F_S(s))g_S(s)\mathrm{d}s, \end{aligned}$$

while the objective function is

$$\begin{aligned} \int_0^{\infty} c\mathrm{d}\Psi(H(c)) &= \int_0^{\infty} H^{[-1]}(F_S(s))\mathrm{d}\Psi(F_S(s)) \\ &= \int_0^{\infty} H^{[-1]}(F_S(s))\Psi'(F_S(s))f_S(s)\mathrm{d}s. \end{aligned}$$

Now make the change of variable $u = F_S(s)$, and write the objective as

$$\int_0^1 H^{[-1]}(u)\Psi'(u)\mathrm{d}u,$$

while the constraint is

$$W = \int_0^1 H^{[-1]}(u)g_S(F_S^{[-1]}(u))\mathrm{d}F_S^{[-1]}(u) = \int_0^1 H^{[-1]}(u)\frac{g_S(F_S^{[-1]}(u))}{f_S(F_S^{[-1]}(u))}\mathrm{d}u.$$

We now write $K(u) = H^{[-1]}(u)$ and $k(u) = K'(u)$, then using integration by parts one gets

$$\int_0^1 (1 - \Psi(u))k(u)\mathrm{d}u$$

for the objective and

$$W = \int_0^1 \left(1 - \int_0^u \frac{g_S(F_S^{[-1]}(v))}{f_S(F_S^{[-1]}(v))}\mathrm{d}v\right) k(u)\mathrm{d}u = \int_0^1 \left(1 - G_S(F_S^{[-1]}(u))\right) k(u)\mathrm{d}u$$

for the constraint.

The aim is finding a $k(u)$ that maximizes

$$\int_0^1 (1 - \Psi(u))k(u)\mathrm{d}u - \alpha \int_0^1 \left(1 - G_S(F_S^{[-1]}(u))\right) k(u)\mathrm{d}u.$$

This function is linear in the unknown function $k(u)$; we have a linear objective and a linear constraint. There are no interior solutions when objectives and constraints are linear in the unknown function. We have here, therefore, a so-called bang-bang solution. For a maximization, if the derivative is positive, one takes the upper bound and if the derivative is negative, one takes the lower bound. The lower bound is zero, since $k(u)$ is the derivative of a non-decreasing inverse of a distribution function. Assume we have a unit finite upper bound for $k(u)$. If one does not assume an upper bound, there is no solution since the problem is linear. We take the upper bound equal to one, since the cash flow is actually only determined up to scale to begin with, and without loss of generality the level of the upper bound can be set to unity. Therefore,

$$k(u) = 1 \text{ if } 1 - \Psi(u) > \alpha \left(1 - G_S(F_S^{[-1]}(u))\right)$$

and zero otherwise.

Hence,

$$K(u) = \int_0^u \mathbf{1}\left[1 - \Psi(v) > \alpha \left(1 - G_S(F_S^{[-1]}(v))\right)\right] \mathrm{d}v,$$

where $\mathbf{1}[A] = 1$ if A is true and zero otherwise.

11

Conic Trading

In this final chapter we show how the conic finance machinery can be put to work to construct optimal trading policies (see also Madan et al. (2016a)). We work in a discrete time Markovian context. The specific activity considered is a simple one of linear positioning in a state-dependent manner on revenues that are correlated with the state transition. Hence one is positioned either to gain or lose the state transition. One may be viewed as going long or short the state transition and hence one is trading the Markov state variable.

We construct an optimal policy or trading strategy by maximizing a conic valuation objective that is the non-linear expectation for the risks undertaken. The aim is to implement such conservative valuations in a context that simultaneously permits the selection of revenue-generating actions that maximize this value.

In order to obtain tractable solutions, we will position ourselves in a steady state situation. The steady state solution is obtained using classical value and policy iteration schemes employed in the literature on Markov decision processes. However, it turns out to be critical here that the valuation is conducted by an expectation with respect to a non-additive probability, for with a classical conditional expectation operator both iterations fail. Since the valuation involves the use of non-linear conditional expectations as they arise in the conic finance modelling of two price economies, we refer to this as *conic trading* (in Markovian steady states). Hence we have a situation in which conic finance actually delivers solutions not available to one-price theories.

Much is written about algorithmic trading and the development of different and more sophisticated prediction algorithms, but they do not simultaneously incorporate the effect of the existing uncertainty around the prediction into the valuation of the strategy. In this chapter we provide a useful approach not only to evaluate prediction algorithms but also to include in the evaluation the effects of residual uncertainties.

We learn how the prediction benefits interact with the interests of hedging the underlying value function for strategies that are to be implemented in a stable environment in perpetuity. From a practical viewpoint we note that algorithms implemented at a one-minute tenor for a day, can essentially be seen as perpetual for Markovian systems, with convergence periods well below such magnitudes. Therefore, system stability for a single day can be long enough for the use of steady state conic trading methods in their evaluation.

11.1 Trading in a Markovian Context

Consider a discrete time Markov process $(x_t, 0 \leq t \leq T)$ describing the state of a system. The process $(x_t, 0 \leq t \leq T)$, for example, could be a particular mean-reverting process which we would like to trade.

For a discrete time Markovian context we have a state transition equation

$$x_{t+1} = A(x_t, t) + u_{t+1}, \qquad t = 0, 1, \ldots.$$

Essentially the next value of our system x_{t+1} is a deterministic (possible time-dependent) function $A(x_t, t)$ of the value in the previous state (x_t) plus a random variable u_{t+1}. Here $E[u_{t+1}] = 0$.

A simple example is a linear system

$$x_{t+1} = a + bx_t + u_{t+1}, \qquad t = 0, 1, \ldots,$$

where a, b are constants. Over time, the process tends to drift towards its long-term mean or the equilibrium state. Such a process is therefore often called *mean-reverting*. If $-1 < b < 1$, such systems are stable and have a limiting stationary distribution. The equilibrium is then given by

$$x_e = a(1 - b)^{-1}.$$

In the sense that, in the absence of the random factor, if the system is currently in this point then

$$x_{t+1} = a + bx_e = a + ab(1 - b)^{-1} = a(1 - b)^{-1} = x_e;$$

hence, we stay in that point.

Example 11.1 *In Figure 11.1, we depict a realization of a linear system with $a =$ 0.05, $b = 0.95$ and u_{t+1} zero-mean normally distributed random variables with standard deviation equal to 4%. Such linear systems driven by normally distributed factors are often termed Ornstein–Uhlenbeck processes. An Ornstein–Uhlenbeck process can also be considered as the continuous-time analogue of the discrete-time autoregressive process. For our parameter setting, the equilibrium point is 1.*

How much we trade, i.e. our policy function, is to be determined and will be in some sense optimized. In general, these policy functions and their valuation can depend on calendar time and the entire information to date. In a Markovian context

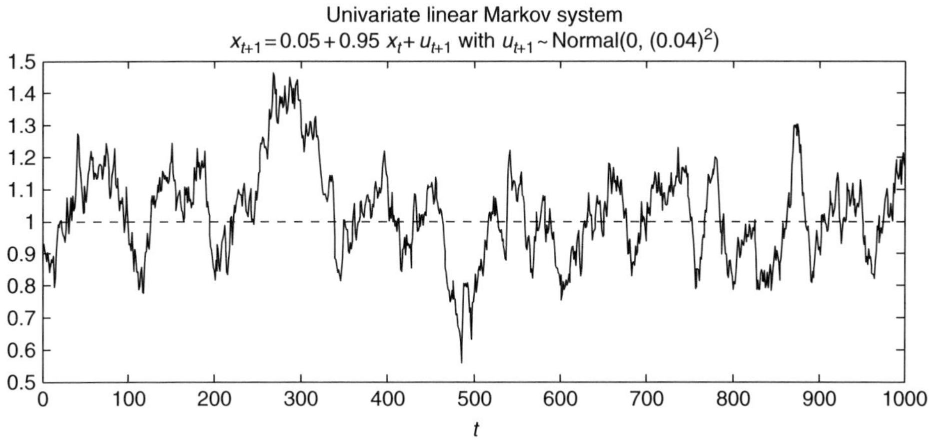

Figure 11.1 Mean-reverting Markov system

the dependence may be reduced to a knowledge of the current state, but both – policies and their values – will still depend on time given a finite horizon for the activity.

We hence deal with policy functions $\alpha(x_t, t)$ taking some actions at time t when the system is in state x_t, to produce the possibly time-dependent revenue at time $t+1$ of $c_{t+1} = c\,(\alpha(x_t, t), x_{t+1}, t+1)$. One should read $\alpha(x_t, t)$ as the amount one is going to trade at time t in the underlier x_t and c_{t+1} is then the cash flow of this trade received at time $t+1$.

In our univariate linear system, we could decide, for example, to take at time t a position of $\alpha(x_t, t)$ in the underlier x_t, leading to a (random) cash flow of

$$\alpha(x_t, t)(x_{t+1} - x_t) = \alpha(x_t, t)(a + (b-1)x_t + u_{t+1}).$$

Note that in equilibrium state, the cash flow is just $\alpha(x_t, t)u_{t+1}$ and has a zero expectation.

11.2 Conic Trading Policy and Value Optimization

First we consider trading from time 0 to time $T-1$, generating the revenue stream $c_1, c_2, \cdots, c_T$. Note that for algorithms implemented intraday with trading, say, every minute or second, discounting doesn't make sense and we hence do not include such discounting; these cash flows are hence just in units of a store of value asset (see also Madan (2016c)).

The conservative value for this activity is given by the non-linear conditional expectation using a particular distortion function Ψ of the stream of cash flows, conditioned on the filtration of the Markov process. Let $V(x, t)$ be the value at time t of the remaining uncertainty being at time t in state x. More precisely, we construct the function $V(x, t)$ as follows:

$$V(x_T, T) = 0 \tag{11.1}$$

$$V(x_t, t) = \mathcal{D}\left[c_{t+1} + V(x_{t+1}, t+1)|\mathcal{F}_t\right] \text{ for } 0 \leq t < T, \tag{11.2}$$

where $c_{t+1} = c\,(\alpha(x_t, t), x_{t+1}, t+1)$ and $\mathcal{D} = \mathcal{D}^{\Psi}$ is used to denote the non-linear or Choquet expectation operator (see Section 4.5). Note that this non-linear expectation is based on information available at time t; in the notation this is made clear by using the standard notation $\mathcal{F}_t$ referring to the information available at time t.

Given a set of feasible policies $\mathcal{G}(x_t, t)$ and a valuation function $V(x, t)$, optimal policies are defined by

$$\alpha^*(x_t, t) = \arg \sup_{\alpha(x_t,t)\in\mathcal{G}(x_t,t)} \mathcal{D}\left[c(\alpha(x_t, t), x_{t+1}, t+1) + V(x_{t+1}, t+1)|\mathcal{F}_t\right]. \tag{11.3}$$

So if we start with an initial candidate policy function $\alpha(x, t)$, then by using Equations (11.1)–(11.2), we can derive a value function $V(x, t)$. Further, given a value function $V(x, t)$ we can obtain using Equation (11.3) an optimal policy function $\alpha^*(x, t)$. For this optimal policy function we again can calculate (using Equations (11.1)–(11.2)) a value function $V^*(x, t)$.

This gives a map from sequences $[\alpha(x, t), V(x, t)]$ to $[\alpha^*(x, t), V^*(x, t)]$. The solution we are after is a fixed point of this mapping, in the sense that we then have the optimal policy with its corresponding value function.

In the interests of tractable solutions, one may implement and analyse the problem in the steady state of the Markov process and in perpetuity ($T \to \infty$) so as to eliminate the dependence on calendar time. Since in such a setting our policy function and our valuation function does not depend on calendar time, we will use the notation $\alpha(x)$ for the policy function and $V(x)$ for the value function. For such a solution, Equations (11.1)–(11.2) simplify and the value functions must satisfy

$$V(x_t) = \mathcal{D}\left[c(\alpha(x_t), x_{t+1}) + V(x_{t+1})|\mathcal{F}_t\right]. \tag{11.4}$$

Likewise, the policy functions satisfy

$$\alpha^*(x_t) = \arg \sup_{\alpha(x_t)\in\mathcal{G}(x_t)} \mathcal{D}\left[c(\alpha(x_t), x_{t+1}) + V(x_{t+1})|\mathcal{F}_t\right]. \tag{11.5}$$

In the current context, x_{t+1} is independent of our policies and hence if the expectation with respect to non-additive probability is replaced by a classical linear expectation then the value function must satisfy

$$V(x_t) = E\left[c(\alpha(x_t), x_{t+1})|\mathcal{F}_t\right] + E[V(x_{t+1})|\mathcal{F}_t];$$

in consequence we get that

$$V(x_t) = E\left[\sum_{j=t+1}^{\infty} c(\alpha\left(x_{j-1}), x_j\right)|\mathcal{F}_t\right] + E[V(x_\infty)|\mathcal{F}_t].$$

The difficulty here is that the infinite sum is stationary and will not be convergent for non-trivial (i.e. non-zero) policies. Hence, the use of classical expectations renders the problem unsolvable. We shall observe that for a distortion there is a solution and the use of non-additive probability is critical. Also, for policy function, if one employs a traditional expectation function then

$$E\left[c(\alpha(x_t), x_{t+1}) + V(x_{t+1})|\mathcal{F}_t\right] = E[c(\alpha(x_t), x_{t+1})|\mathcal{F}_t] + E[V(x_{t+1})|\mathcal{F}_t]$$

and by linearity of c in α there will be no interior solution for the arg sup. But a non-additive probability will account for the diversification between

$$c\,(\alpha(x_t), x_{t+1}) \text{ and } V(x_{t+1})$$

and delivers an interior solution for

$$\arg\sup_{\alpha(x)} \mathcal{D}\left[c(\alpha(x_t), x_{t+1}) + V(x_{t+1})|\mathcal{F}_t\right]$$

even when c is linear.

There are two implementation details to be addressed before one may exhibit an implementation. It is clear from the structure of the value function recursion that one may add an arbitrary constant to the value function as it will remain a value function. Hence one may set the value function to be zero at the equilibrium point, for example, by just subtracting this value from any candidate where it is not zero. It is also clear that the value function may be scaled by a positive constant by similarly scaling the policy. Hence the value function is also determined up to scale. To avoid the identically zero solution we scale the value function to be ± 1 at an extreme point above the equilibrium point. Scaling by a positive constant does not permit one to change the sign at such an upper point, but we can preserve sign and scale to a positive or negative unity. With these adjustments for level and scale we implement the value and policy iteration in Equations (11.4) and (11.5) to convergence to a fixed-point solution.

Example 11.2 *We recall the setting of Example 11.1 with dynamics:*

$$x_{t+1} = .05 + .95x_t + u_{t+1}$$

where the u_t, $t = 1, 2, \ldots.$ are independent and identically distributed following a zero-mean normal distribution with a standard deviation of 4%. We use the MINMAXVAR distortion with distortion parameter $\lambda = 0.5$. The convergence was accomplished in 50 iterations. Furthermore, calculations are performed on a non-uniform state space grid of size 100. Figures 11.2 and 11.3 present respectively the value and policy function solutions.

We observe from the solution that near equilibrium the force of the drift is small and the value function reflects a local minimum at the equilibrium point. Away from

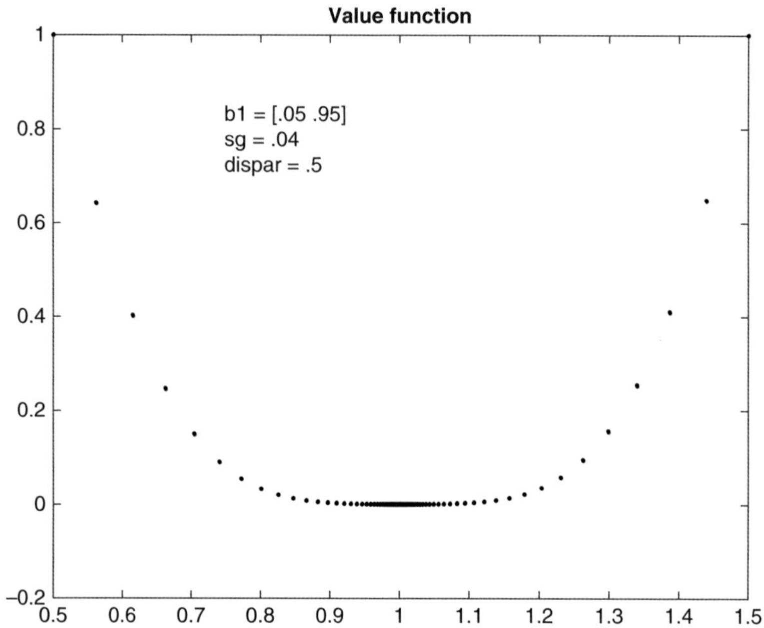

Figure 11.2 Value function for indicated parameter settings

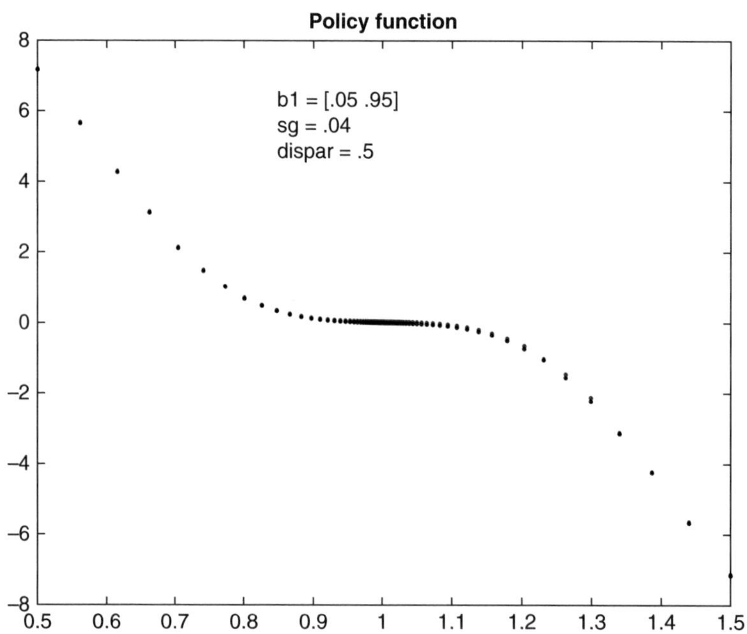

Figure 11.3 Policy function solution

Table 11.1 *Four cases with regard to momentum and mean reversion.*

	M increasing in x	M decreasing in x
G increasing in x	MMD, MRU	MMD, MMU
G decreasing in x	MRD, MRU	MRD, MMU

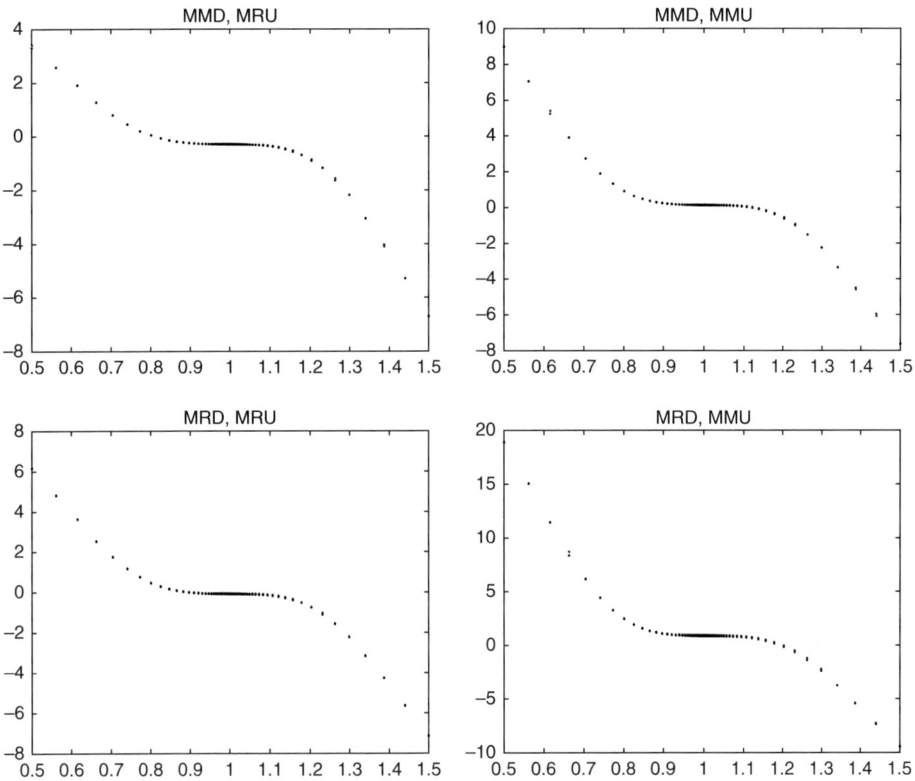

Figure 11.4 Four optimal policy function

equilibrium the drift picks up and the value function rises on both sides, inducing the policy to go short/long on account of both the hedge for the value function and drift considerations when above/below the equilibrium.

Example 11.3 *For a further investigation of how the policy function responds to the structure of the residual component we consider variance gamma (Madan and Seneta (1990), Madan, Carr and Chang (1998)) residuals with a base case in the CGM (Carr, Geman, Madan and Yor (2002)) parametrization of* $C = 3$, $G = 70$

and $M = 60$. *We may next let the* G, M *parameters vary linearly with the state variable* x. *If* M *increases with* x *then arrival rates of up moves fall as* x *rises and we have mean reversion up (MRU). Similarly, if* M *decreases with* x *we have up moves rising as* x *rises or momentum up (MMU). Likewise,* G *falling with* x *gives mean reversion down (MRD) while a positive relationship between* G *and* x *represents momentum down (MMD). With slope coefficients of* ± 10 *for* G, M *we have four cases with regard to momentum and mean reversion up or down summarized in Table 11.1. We present four graphs for the four policy functions in Figure 11.4. We observe that mean reversion in either direction enhances the mean reversion play.*

References

Albrecher, H., Guillaume, F. and Schoutens, W. (2013). Implied liquidity: model sensitivity. *Journal of Empirical Finance* **23**, 48–67.

Artzner, P., Delbaen, F., Eber, J.-M. and Heath, D. (1999). Coherent measures of risk. *Math. Finance* **9** (3), 203–28.

Bachelier, L. (1900). Théorie de la spéculation. *Ann. Sci. Ecole. Norm. Sup.* **17**, 21–86.

Bakshi, G. and Madan, D.B. (2000). Spanning and derivative security valuation. *Financial Economics* **55**, 205–38.

Barndorff-Nielsen, O.E. (1977). Exponentially decreasing distributions for the logarithm of particle size. *Proceedings of the Royal Society London A* **353**, 401–19.

Barndorff-Nielsen, O.E. (1978). Hyperbolic distributions and distributions on hyperbolae. *Scandinavian Journal of Statistics* **5**, 151–7.

Barndorff-Nielsen, O.E. (1995). *Normal inverse Gaussian distributions and the modeling of stock returns*. Research Report No. 300, Department of Theoretical Statistics, Aarhus University.

Barndorff-Nielsen, O.E. (1996). Normal inverse Gaussian distributions and stochastic volatility models. *Scandinavian Journal of Statistics* **24**, 1–13.

Barndorff-Nielsen, O.E. (1998). Processes of normal inverse Gaussian type. *Finance and Stochastics* **2**, 41–68.

Barndorff-Nielsen, O.E. and Halgreen, C. (1977). Infinitely divisibility of the hyperbolic and generalized inverse Gaussian distributions. *Zeitschrift für Wahrscheinlichkeitstheorie und verwandte Gebiete* **38**, 309–11.

Bates, D. (1996). Jump and stochastic volatility: exchange rate processes implicit in Deutsche Mark options. *Reviews of Financial Studies* **9**, 69–107.

Black, F. and Scholes, M. (1973). The pricing of options and corporate liabilities. *Journal of Political Economy* **81**, 637–54.

Carr, P. and Madan, D.B. (1998). Option valuation using the fast Fourier transform. *Journal of Computational Finance* **2**, 61–73.

Carr, P. and Madan, D.B. (2001). Towards a theory of volatility trading. In *Option Pricing, Interest Rates, and Risk Management* Jouini, E., Cvitanic, J. and Musiela, M. (eds). Cambridge University Press, 458–76.

Carr, P. and Madan, D.B. (2005). A note on sufficient conditions for no arbitrage. *Finance Research Letters* **2**, 125–30.

Carr, P. and Madan, D.B. (2013). Joint modeling of VIX and SPX options at a single and common maturity with risk management applications. *Transactions of the Institute of Industrial Engineers* **46**, 1125–31.

Carr, P., Madan, D.B. and Vicente Alvarez, J.J. (2011). Markets, profits, capital, leverage and return. *Journal of Risk* **14**, 95–122.

Carr, P., Geman, H., Madan, D.H. and Yor, M. (2002). The fine structure of asset returns: an empirical investigation. *Journal of Business* **75**, 305–32.

Carr, P., Geman, H., Madan, D.B. and Yor, M. (2003). Stochastic volatility for Lévy processes. *Mathematical Finance*, Vol. 13, No. 3, 345–82.

Carr, P., Madan, D.B., Melamed, M. and Schoutens, W. (2016). Hedging insurance books. *Insurance: Mathematics and Economics*, forthcoming.

Cherny, A. and Madan, D.B. (2009). New measures for performance evaluation. *Review of Financial Studies* **22**, 2571–606.

Cherny, A. and Madan, D.B. (2010). Markets as a counterparty: an introduction to conic finance. *International Journal of Theoretical and Applied Finance* **13**, 1149–77.

Choquet, G. (1953). Theory of capacities. *Ann. Inst. Fourier, Grenoble* **5**, 131–295.

Corcuera, J., Guillaume, F., Leoni, P. and Schoutens, W. (2009). Implied Lévy volatility. *Quantitative Finance* **9** (4), 383–93.

Corcuera, J.M., Guillaume, F., Madan, D.B. and Schoutens, W. (2012). Implied liquidity: towards stochastic liquidity modelling and liquidity trading. *International Journal of Portfolio Analysis and Management* **1** (1), 80–91.

Cox, J., Ingersoll, J. and Ross, S. (1985). A theory of the term structure of interest rates. *Econometrica* **53**, 385–408.

Delbaen, F. (2002). *Coherent Risk Measures on General Probability Spaces.* In *Advances in Finance and Stochastics Essays in Honour of Dieter Sondermann* Klaus Sandmann and Philipp J. Schönbucher (eds). Springer, 1–37.

Derman, E., Kani, I. and Chriss, N. (1996). Implied trinomial trees of the volatility smile. *Quantitative Strategies Research Notes*, Goldman Sachs, February 1996.

Devroye, L. (1986). *Non-Uniform Random Variate Generation.* New York: Springer-Verlag.

Eberlein, E. and Jacod, J. (1997). On the range of options prices. *Finance and Stochastics* **1** (2), 131–40.

Eberlein, E. and Keller, U. (1995). Hyperbolic distributions in finance. *Bernoulli* **1**, 281–99.

Eberlein, E. and Madan, D.B. (2009). Hedge fund performance: sources and measures. *International Journal of Theoretical and Applied Finance* **12**, 267–82.

Eberlein, E. and Madan, D.B. (2012). Unbounded liabilities, capital reserve requirements and the tax payer put option. *Quantitative Finance* **12** (5), 709–24.

Eberlein, E. and Madan, D.B. (2014). Maximally acceptable portfolios. In *Inspired by Finance: A Festschrift for Marek Musiela* Kabanov, Y., Rutkowski, M. and Zariphopolou, T. (eds). Berlin: Springer-Verlag, 257–72.

Eberlein, E. and Madan, D.B. (2015). Portfolio theory for squared returns correlated across time. *Working Paper*, Robert H. Smith School of Business (http://ssrn.com/abstract=2635632).

Eberlein, E., Gehrig, T. and Madan, D.B. (2012). Pricing to acceptability: with applications to the pricing of one's own credit risk. *Journal of Risk* **15** (1), 91–120.

Eberlein, E., Madan, D.B. and Schoutens, W. (2013). Modeling risk weighted assets and the risk sensitivity of related capital requirements. *Journal of Risk* **16**, 2.

Eberlein, E., Madan, D.B., Pistorius, M. and Yor, M. (2014a). Bid and ask prices as non-linear continuous time G-expectations based on distortions. *Mathematics and Financial Economics* **8**, 265–89.

Eberlein, E., Madan, D.B., Pistorius, M., Schoutens, W. and Yor, M. (2014b). Two price economies in continuous time. *Annals of Finance* **10**, 71–100.

Fu, M.C., Jarrow, R.A., Yen, J.-Y.J. and Elliott, R.J. (2000). *Advances in Mathematical Finance*. Boston: Birkhäuser.

Geman, H., Madan, D. and Yor, M. (2001). Time changes for Lévy processes. *Mathematical Finance* **11**, 79–96.

Grigelionis, B. (1999). Processes of Meixner type. *Lithuanian Mathematical Journal* **39** (1), 33–41.

Guillaume, F. (2011). *Applications of Lévy Processes in Credit and Volatility Modelling*. PhD manuscript. KU Leuven and TU Delft.

Guillaume, F. (2015). The LIX: a model-independent liquidity index. *Journal of Banking and Finance* **58**, 214–31.

Guillaume, F. and Schoutens, W. (2013a). A moment matching market implied calibration. *Quantitative Finance* **13** (9), 1359–73.

Guillaume, F. and Schoutens, W. (2013b). Bid–ask spread for exotic options under conic finance. In *Innovations in Quantitative Risk Management* Glau, K., Scherer, M. and Zagst, R. (eds). New York: Springer, 59–74.

Halgreen, C. (1979). Self-decomposability of the generalized inverse Gaussian and hyperbolic distributions. *Zeitschrift für Wahrscheinlichkeitstheorie und verwandte Gebiete* **47**, 13–18.

Heston, S.L. (1993). A closed form solution for options with stochastic volatility with applications to bonds and currency options. *Review of Financial Studies* **6** (2), 327–43.

Kusuoka, S. (2001). On law invariant coherent risk measures. *Advances in Mathematical Economics* **3**, 83–95.

Lamperti, J. (1962). Semi-stable stochastic processes. *Transactions of the American Mathematical Society* **104**, 62–78.

Madan, D.B. (2009). Capital requirements, acceptable risks and profits. *Quantitative Finance* **9**, 767–73.

Madan, D.B. (2010a). Pricing and hedging basket options to a prespecified level of acceptability. *Quantitative Finance* **10**, 607–15.

Madan, D.B. (2010b). Conserving capital by adjusting deltas for gammas in the presence of skewness. *Journal of Risk and Financial Management* **3**, 1–25.

Madan, D.B. (2011). Joint risk neutral laws and hedging. *Transactions of the Institute of Industrial Engineers* **43**, 840–50.

Madan, D.B. (2012a). Execution costs and efficient execution frontiers. *Annals of Financial Economics* **7** (1). DOI: 10.1142/S2010495212500029.

Madan, D.B. (2012b). A two price theory of financial equilibrium with risk management implications. *Annals of Finance* **8** (4), 489–505.

Madan, D.B. (2012c). From credit valuation adjustments to credit capital commitments. *Quantitative Finance* **12** (6), 839–45.

Madan, D.B. (2014). Modeling and monitoring risk acceptability in markets: the case of the credit default swap market. *Journal of Banking and Finance* **47**, 63–73.

Madan, D.B. (2015). Asset pricing theory for two price economies. *Annals of Finance* **11**, 1–35.

Madan, D.B. (2016a). Conic portfolio theory. *International Journal of Theoretical and Applied Finance* **19**, 1650019. DOI: 10.1142/S0219024916500199.

Madan, D.B. (2016b). Marking to two price markets. *Journal of Asset Management* **17**, 100–18.

Madan, D.B. (2016c). Relativities in financial markets. *Working Paper*, Robert H. Smith School of Business (http://ssrn.com/abstract=2732825).

Madan, D.B. and Milne, F. (1991). Option pricing with V.G. martingale components. *Mathematical Finance* **1** (4), 39–55.

Madan, D.B. and Schoutens, W. (2011a). Conic coconuts: the pricing of contingent capital notes using conic finance. *Mathematics and Financial Economics* **4**, 87–106.

Madan, D.B. and Schoutens, W. (2011b). Conic financial markets and corporate balance sheet. *International Journal of Theoretical and Applied Finance* **14**, 587–610.

Madan, D.B. and Schoutens, W. (2012a). Structured products equilibria in conic two price markets. *Mathematics and Financial Economics* **6** (1), 37–57.

Madan, D.B. and Schoutens, W. (2012b). Tenor specific pricing. *International Journal of Theoretical and Applied Finance* **15** (6). DOI:10.1142/S0219024912500434.

Madan, D.B. and Schoutens, W. (2013). Systemic risk tradeoffs and option prices. *Insurance, Mathematics and Economics* **52**, 222–30.

Madan, D.B. and Schoutens, W. (2014). Two processes for two prices. *International Journal of Theoretical and Applied Finance* **17** (1). DOI: 10.1142/S0219024914500058.

Madan, D.B. and Schoutens, W. (2016). Conic CVA and DVA. *Willmott Magazine*, Issue 82, pages 45–50.

Madan, D.B. and Seneta, E. (1987). Chebyshev polynomial approximations and characteristic function estimation. *Journal of the Royal Statistical Society Series B* **49** (2), 163–9.

Madan, D.B. and Seneta, E. (1990). The V.G. model for share market returns. *Journal of Business* **63**, 511–24.

Madan, D.B. and Sharaiha, Y. (2015). Option overlay strategies. *Quantitative Finance* **15**, 1175–90. DOI: 10.1080/14697688.2014.989898.

Madan, D.B., Carr, P. and Chang, E.C. (1998). The variance gamma process and option pricing. *European Finance Review* **2**, 79–105.

Madan, D.B., Pistorius, M. and Schoutens, W. (2013). The valuation of structured products using Markov chain models. *Quantitative Finance* **13** (1), 125–36.

Madan, D.B., Pistorius, M. and Schoutens, W. (2016a). Conic trading of a markov process. *Working Paper*, Robert H. Smith School of Business (http://ssrn.com/abstract=2732826).

Madan, D.B., Pistorius, M. and Schoutens, W. (2016b). Dynamic conic hedging for competitiveness. *Mathematics and Financial Economics*, 1–35.

Madan, D.B., Pistorius, M. and Stadje, M. (2016). Convergence of BSDEs driven by random walks to BSDEs: the case of (in)finite activity jumps with general driver. *Stochastic Processes and their Applications* **126**(5), 1553–84.

Maller, R.A., Solomon, D.H. and Szimayer, A. (2006). A multinomial approximation of American option prices in a Lévy process model. *Mathematical Finance* **16** (4), 613–33.

Merton, R.C. (1973). Theory of rational option pricing. *Bell J. Econom. Managem. Sci.* **4**, 141–83.

Samuelson, P. (1965). Rational theory of warrant pricing. *Industrial Management Review* **6**, 13–32.

Sato, K. (2000). *Lévy Processes and Infinitely Divisible Distributions*. Cambridge Studies in Advanced Mathematics **68**. Cambridge University Press.

Sato, K. and Yamazato, M. (1982). Stationary processes of Ornstein–Uhlenbeck type. In *Probability Theory and Mathematical Statistics* Itô, K. and Prohorov, J.V. (eds). Lecture Notes in Mathematics **1021**. Berlin: Springer.

Schoutens, W. (2001). *The Meixner Process in Finance*. EURANDOM Report 2001-002, Eindhoven: EURANDOM.

Schoutens, W. (2002). *Meixner Processes: Theory and Applications in Finance*. EURANDOM Report 2002-004, Eindhoven: EURANDOM.

Schoutens, W. (2003). *Lévy Processes in Finance: Pricing Financial Derivatives*. New York: Wiley.

Wang, S.S. (1995). Insurance pricing and increased limits ratemaking by proportional hazards transforms. *Insurance: Mathematics and Economics* **17**, 43–54.

Wang, S.S. (2000). A class of distortion operators for pricing financial and insurance risks. *Journal of Risk and Insurance* **67**, 15–36.

Wang, S.S. (2002a). A universal framework for pricing financial and insurance risks. *Astin Bulletin* **32**, 213–34.

Wang, S.S. (2002b). Equilibrium pricing transforms: new results using Bühlmann's 1980 economic model. *Astin Bulletin* **33**, 57–73.

Yamada, Y. and Primbs, J.A. (2004). Properties of multinomial lattices with cumulants for option pricing and hedging. *Asia-Pacific Financial Markets* **11**, 335–65.

Index